Above
the Arctic Circle

The Alaska Journals
of James A. Carroll
1911 - 1922

by James A Carroll

Publication Consultants Since 1978

PO Box 221974 Anchorage, Alaska 99522-1974

ALASKA

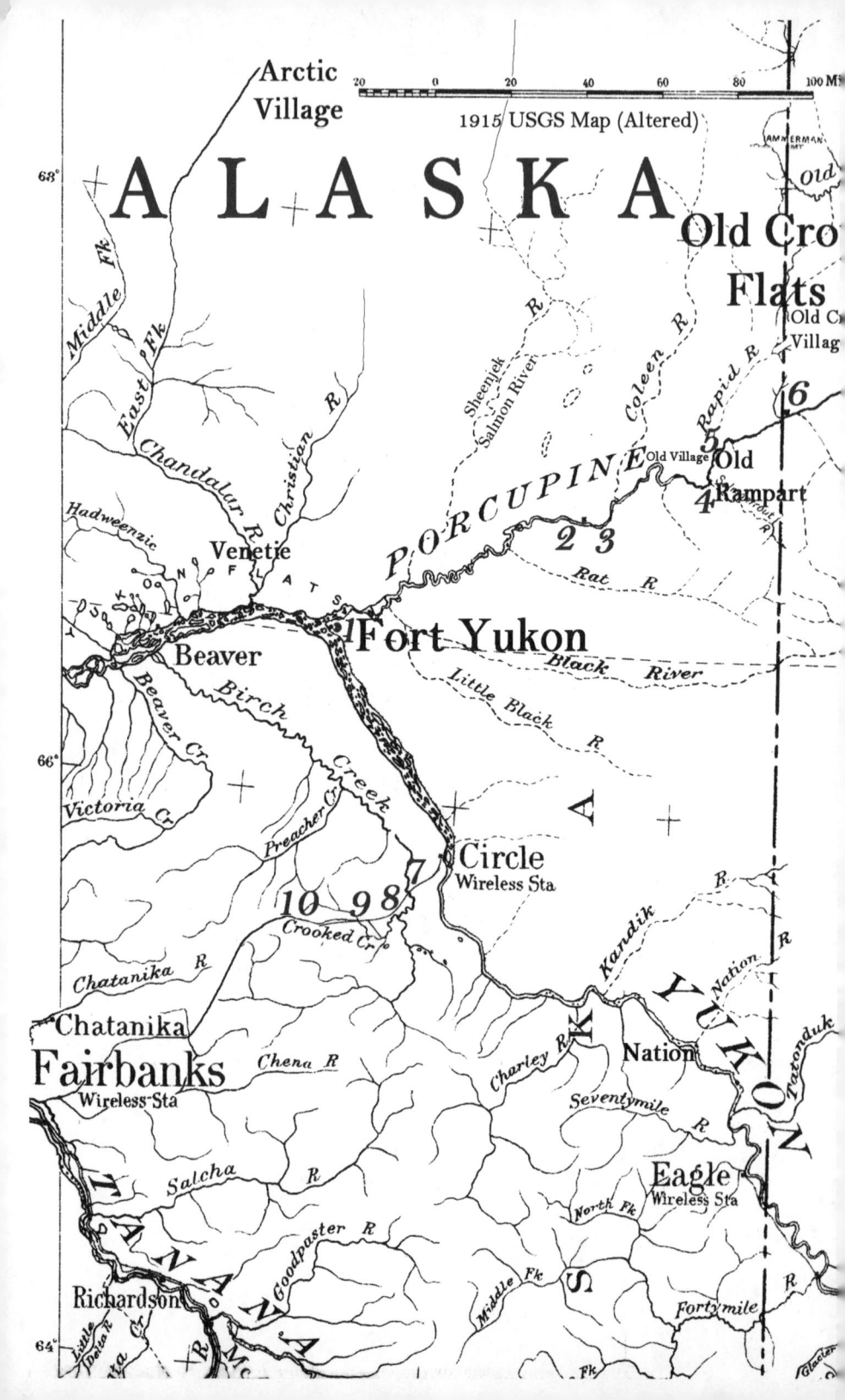

Arctic Village

1915 USGS Map (Altered)

Old Crow Flats

Old Crow Village

Middle Fk
East Fk
Chandalar R.
Hadweenzic
Christian R.
Sheenjek R.
Salmon River
Coleen R.
Rapid R.
Old Village
Old Rampart
Venetie
PORCUPINE
Rat R.
FLATS
Beaver
Fort Yukon
Beaver Cr.
Birch Creek
Little Black R.
Black River
Victoria Cr.
Preacher C.
Circle Wireless Sta.
Crooked Cr.
Kandik R.
Nation R.
Chatanika R.
Chatanika
Fairbanks Wireless Sta.
Chena R.
Charley R.
Nation
Seventymile R.
Tatonduk
Salcha R.
Eagle Wireless Sta.
North Fk
TANANA
Richardson
Goodpaster R.
Middle Fk
Fortymile R.
Glacier
YUKON

Ammerman Mt.

ROADHOUSES

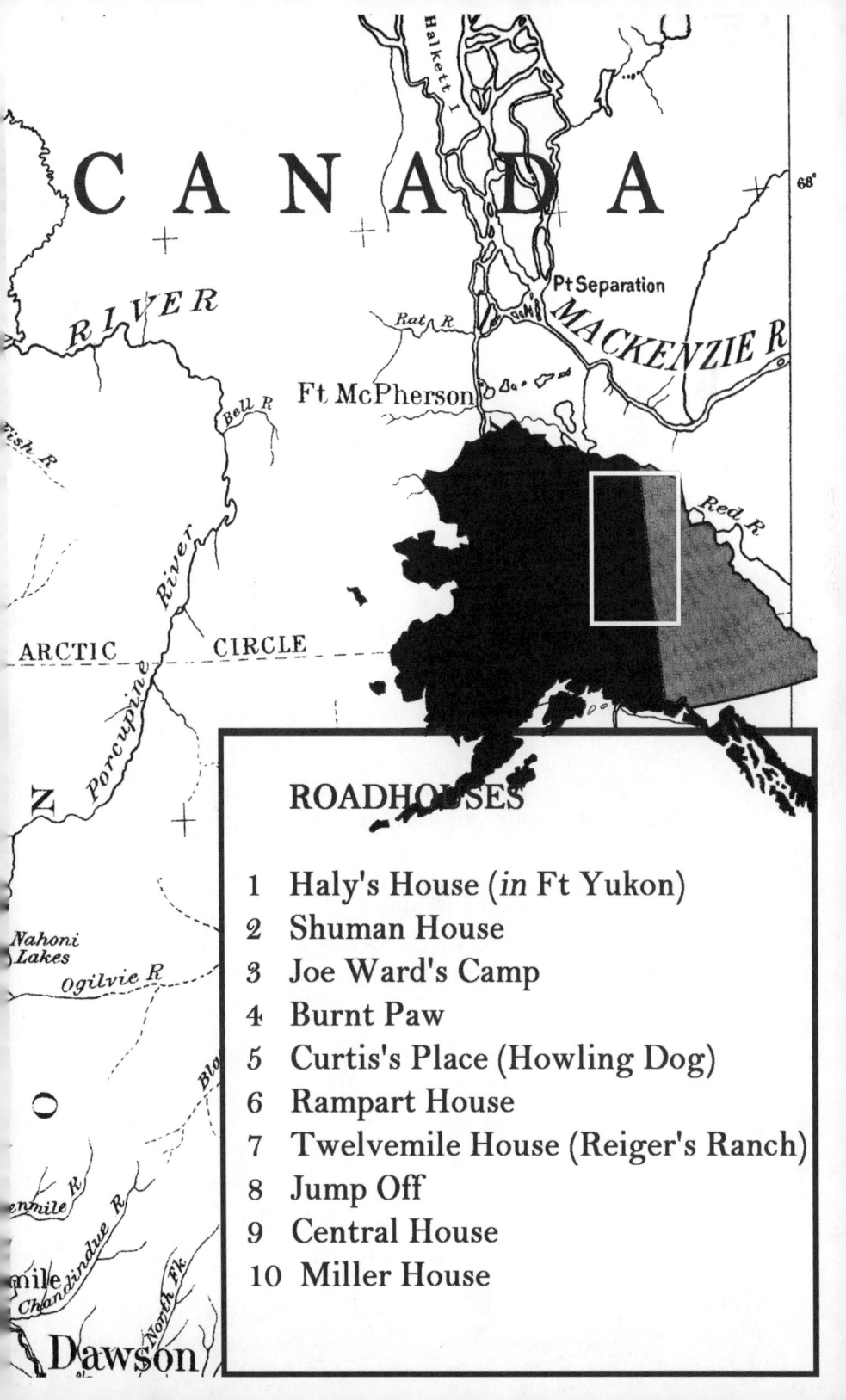

1. Haly's House (*in* Ft Yukon)
2. Shuman House
3. Joe Ward's Camp
4. Burnt Paw
5. Curtis's Place (Howling Dog)
6. Rampart House
7. Twelvemile House (Reiger's Ranch)
8. Jump Off
9. Central House
10. Miller House

ISBN 1-59433-032-8

Library of Congress Catalog Card Number: 2005904546

©1957 Copyright by James A. Carroll. First Edition "The First Ten Years in Alaska – Memoirs of a Fort Yukon Trapper 1911 – 1922" published by Exposition Press, Inc.

©2005 Copyright by Lorraine Kirker. Second Edition. Published by Publication Consultants.

All rights reserved, including the right of reproduction in any form, or by any mechanical or electronic means including photocopying or recording, or by any information storage or retrieval system, in whole or in part in any form, and in any case not without the written permission of the author and publisher.

Introduction, epilogue, and cover designs by
Cheryl and Joe Homme

Manufactured in the United States of America.

Dedication

To all my grandchildren

Contents

Map 4

Dedication 7

Acknowledgments 11

Introduction to the Second Edition 13

CHAPTER I .... North of the Arctic Circle 15

CHAPTER II .... Wintering in Circle City 31

CHAPTER III ...... Back on the Trap Line 47

CHAPTER IV ...... Pulling Out 63

CHAPTER V .... Trapping With a Family 83

CHAPTER VI ...... The Crow Flats 105

CHAPTER VII ..... Back at Fort Yukon 123

CHAPTER VIII ...... The K-Brothers' Strange Disappearance 137

CHAPTER IX ...... Home at Last 147

Epilogue to the Second Edition 155

Other books about Fort Yukon 158

Acknowledgments

Lorraine Kirker wishes to thank the friends, family members, and professionals who helped bring this book to a new generation. Her store, the Alaskana Bookshop in Palmer, carries more than 25,000 books about Alaska. James A. Carroll's "The First Ten Years in Alaska," however, has never graced its shelves. This hard-to-find Alaska classic has languished on a long customer waiting list. Finally, Lorraine acquired a copy and, after reading it, knew this was a book that needed to be reprinted. That decision was the easy part. The process has taken two years with much of the technical work being done by Cheryl Homme. In addition to her, Lorraine would like to thank the following people for their help and support.

Townspeople of Fort Yukon
The family of James A. and Fannie Carroll
Richard Jr. and Kathleen Carroll
Richard Sr. and Eva Carroll
Jennie Carroll
Evelyn James
Candy Waugaman
Anna Buterbaugh
Lynette Lehn
Dennis and Karen Goodenough
Patricia Pauley
Jude Baldwin
Donna Rae Thompson
Laura Wood
Mary Ann Slemmons
Rose Speranza
Family and Friends

Introduction to the Second Edition

In 1910, as a young man of seventeen, James A. Carroll traveled from his hometown in Aitkin, Minnesota to Seattle, where he boarded a steamship bound for Alaska. His older brother had encouraged the trip, having written to James about "easy pickings, gold nuggets the size of walnuts." Within two years of his journey north, James Carroll settled in Fort Yukon, 140 miles north of Fairbanks and 200 river miles from the Canadian border. Surrounded by hundreds of lakes, five major rivers, and centered in the fertile Yukon Flats, Fort Yukon was a close-knit community that welcomed the young James Carroll from the outset. It was from Fort Yukon that he embarked on his far-ranging adventures in the Land of the Midnight Sun. This period he chronicled in a series of journals that first reached print under the title "The First Ten Years in Alaska – Memoirs of a Fort Yukon Trapper 1911 – 1922." This second edition mirrors the text of the 1957 publication with the correction of minor errors that escaped the original editor. It remains the story of a young man initially beckoned to Alaska by the lure of nuggets, but who found it was not the gold that would keep him – instead it was the people, the land, and his family that would sustain him during the next fifty-three years of living above the Arctic Circle.

CHAPTER I
North of the Arctic Circle

When I first came to Alaska in 1910, cooking was the only occupation I knew anything about. In our family, there were eight boys with only one sister. When she got married in 1901, it fell to us boys to do most of the cooking at home. This we did in turns, doing both girls' and boys' jobs, such as ironing our clothes, washing dishes, etc. We raised about everything we ate, including pork and beef, which was cheap. I remember we used to sell steers for five cents a pound on the hoof. We also used to sell eggs for ten cents a dozen.

I got a job in a logging camp north of Aitkin, Minnesota, working for a logger by the name of Fred Blaise, in 1905 – 1906. Blaise's nephew was the cook; I was the "bull" cook. The nephew used to let me do most of the cooking, which was good practice for me. We never had much variety. We had plenty of meats and beans, as well as yellow peas for split-pea soup. There was only a small crew to cook for—about fourteen men. It was also my job to take the crew a hot lunch at noon. I used to haul this lunch on a small sled. It was pretty much the same each day: a pot of baked beans, doughnuts, black molasses, cake, bread and butter. The whole dinner was well wrapped in blankets for warmth. Someone in the crew always had a kettle of boiling water for the tea hanging over the campfire. We all ate lots of deer meat; deer could be seen most every day. I stayed with Blaise until spring. I was paid wages of fifty cents per day.

During the winter of 1906 – 1907 I worked for Blaise again as cook. I handled the cooking job alone. I got a raise in pay of five dollars per month which allowed me twenty dollars per month. One of the crew used to come in at noon to take the hot lunch out as I had done. Blaise's nephew went back to Montreal, which was his home. But 1907 was a year of panic. I was very fortunate to have any kind of a job that year. I always used to have a hard time making my cakes stand up, even with my mother trying to teach me. Finally, after ditching about a dozen cakes that had fallen to the bottom of the cake pan, I reasoned with myself that I must have made my cake batter too thin and rich by adding too much baking powder. After that, I had more success.

After Blaise closed up his logging camp for the season I went home with my hundred and twenty dollars for six months' work. I gave a hundred dollars to my father and kept twenty dollars for myself.

During the next two years I worked for the Weyerhaeuser Lumber Company, cooking for six men. These six men were timber cruisers and estimators for the company which owned vast stands of virgin timber all over northern Minnesota. We were on the go nearly all the time, moving from one section of the country to another, using pack horses and sometimes wagons.

I had to do all the cooking over a campfire. I used a kind of a grate with half-inch round iron legs at each corner. These legs were pushed down in the ground about six inches to hold the grate steady. Over this grate I did my frying and boiling and pastry-cooking. I used a tin reflector oven; this was set up facing a bed of hot coals. I could bake anything with the reflector: pies, biscuits, etc. I used to cook our beans in a bean hole—this bean hole was made by digging a hole in the ground 3 feet by 2 feet. I would fill this hole full of small cut wood. When all the wood burned down to hot coals, I would rake a hole in these hot coals and set the bean pot into the center and rake the hot coals back over the pot; and then I would cover it all over with six or eight inches of dirt. Generally I would do this before going to bed. We would have piping-hot beans for breakfast. In preparing the beans for the hole I used a gallon kettle with a tight-fitting cover. I would fill this kettle a third full of dry beans, salt and pepper, one onion, bacon cut in squares, some black molasses and some tomato sauce; then I filled the pot with water to one-half inch from the top.

I used to cook fish in this way also. All I would put in with the fish would be seasoning and a few strips of bacon. If you cared to, you could eat the bones with the fish – the bones were as soft as the fish. Foods cooked in this way tasted delicious because none of the flavors could escape. The "Big Boss," Charles Weyerhaeuser himself, used to drop in on us and have lunch. He always used to tell me how good everything tasted.

It was now the spring of 1910. I had my mind made up to go to Alaska. Mr. Weyerhaeuser told me I could work for his company if I did not like Alaska and came back again. On my departure he wished me the best of luck and hoped nothing would happen to me up there in far-off Alaska among the Eskimos, igloos and ice.

I landed in Dawson City, Yukon Territory, on August 17, 1910, from my home town, Aitkin. I was seventeen years old and full of ambitious ideas about getting rich quick in Alaska and going back home to enjoy my wealth. This was forty-six years ago—I am still waiting to overtake that pot of gold.

I came into Alaska via Dawson City, over the White Pass and Yukon route. I was compelled to wait in Dawson a full week to make boat connections for downriver points. A handful of tourists were waiting for the same boat. These boats went downstream as far as Saint Michael at the mouth of the Yukon. From there passengers, if they chose, could come back up to Dawson or board a ship for Seattle.

17

During our wait in Dawson we used to make short trips out to the gold diggings or placer mines. Heaps of loose gravel were everywhere. These were tailing piles—gravel that has already been worked out for its gold content. We thought the miner might have overlooked a few nuggets; we used to spend hours sifting this gravel through our fingers in hope of finding some nuggets. We were all called "cheechakos," a name given to all newcomers who have been in Alaska less than a year or have not witnessed the breakup of the Yukon River. Dawson was quite a town, even at that time. Later on, the big dredge companies bought out all the small operators, who in turn left Dawson for the "outside"; this just about turned Dawson into another ghost town.

On our strolls about Dawson we noticed some nice gardens and flower beds. The first frost of the season occurred August 23. Puddles in the streets, caused by recent rain, were skinned over with ice. There is still, to this day, lots of gold being mined in the Dawson and Klondike areas.

The boat we were all waiting for finally arrived. Her name was *Sarah*; she was one of the largest boats to ply the Yukon River at that time. We all climbed aboard and were soon on our way downstream again. The boat made a brief stop at Eagle, Alaska; a few soldiers got off the boat there, and some supplies were dropped. Eagle, at that time, was a military post.

The boat's next stop was at Nation, a small place between Eagle and Circle City. Some freight was put ashore at Nation. At the next stop, Circle, the boat stopped only about half an hour. That was my destination. There must have been forty people lining the riverbank, watching the steamer as it eased its way to shore to be tied up. The audience on shore consisted mostly of women, children, and old men; and, of course, all the businessmen of the town were on hand to greet the captain and the lesser officers. All passengers aboard were allowed half an hour ashore. I happened to be the only passenger getting off the boat at Circle.

I wore a stiff plug hat when I came into Alaska—this type of hat back home was in high style—but in Alaska it was very much out of style. Everybody seemed to be wearing caps of dark colors. I had such a cap in my suitcase. I didn't know what to do with my plug hat. I would look like a hayseed wearing it ashore. I thought of throw-

Fort Yukon viewed from the deck of a ship in 1904. Carroll arrived in Circle, Alaska in 1910 aboard the Sarah. *(PCA 75-428 Alaska State Library Paul Sinic Photograph Collection; C.L. Andrews, photographer)*

ing it overboard in the river, but this wouldn't do, as people seeing a hat floating down the river would think somebody had fallen overboard and drowned. To simplify matters I left the hat in my stateroom and wore my cap ashore. This corresponded better with what the Alaskans were wearing for headgear. But I wasn't through with the hat that easily. After everybody got aboard the boat again and the crew were pulling in the gangplank, the purser came out, rushed up to the railing waving the plug hat and wanting to know who had left his hat aboard the boat. I shouldn't have said anything, but I blurted out that the hat was mine, and that I left it aboard the boat to get rid of it. By this time everybody ashore and on the boat had their eyes focused on the plug hat and me.

There were a few native families living at one end of the town; they were all very friendly people. In those days the natives never lived in town during the summer

Fort Yukon in the early 1900s. Salmon are dried on racks for winter food for consumption by people and dogs. (Courtesy Candy Waugaman)

months, they preferred to live in tents up and down the Yukon River or on smaller streams like Birch Creek. They all fished for salmon with fish wheels, which would catch up to four hundred king salmon each day. There would be two or more families camped at each fish wheel to help with the cutting of the fish. This cutting consisted of slicing the fish lengthwise and crosswise and hanging them on racks to dry.

Some of this dried fish was sold to the store; the best and richest of this fish they kept for their own consumption. The silver salmon they dried, mostly for dog feed.

They did their hunting for meat in the fall when the weather got cool enough for the meat to keep. They used to kill enough moose and caribou to keep them in meat during the cold winter months. Most of the families would have a ton or so of dried fish, fit for man or beast to eat. Most of the natives spent their winters in Circle; they brought in the fish and meat that they took during summer and fall and stored it in their caches for winter use. They sold much of their fish to the U.S. mail carriers for dog feed. The mail was carried by dog teams from Fairbanks to Eagle at that time, and also to Fort Yukon.

I stayed in Circle two days before starting the long walk out to the mine near Miller House, about seventy miles from Circle City. My brother, Tom, was cooking for a man named Pete Anderson. Incidentally, this is the same Anderson who was drowned with his wife when the Canadian steamer, *S.S. Sophia*, hit a reef during a severe storm and all aboard (over three hundred) were lost in 1918. Pete Anderson owned rich placer ground on Mastodon Creek, up about five miles from Miller House. The Andersons used to spend their winters in Seattle. There was only a summer packhorse trail between Circle City and Miller House. In wintertime the freighters used sleds and double-enders, which are light sleds turned up at both ends. This winter trail followed the swamps and lakes most of the way. I walked to Miller House in three days from Circle. Before getting to Miller House I ran across a prospector. He was sinking a shaft in hope of hitting a paying streak of gold, using wood fires to thaw the permafrost. For one man it was a job of climbing up and down a ladder all day. He would sink down about a foot each day. He was sinking the hole at the saw pit where miners had whipsawed sluice box lumber for the mining operations. There were some spruce trees still standing by the old saw-pit site.

I hollered hello to the old prospector; he said hello back.
I said to him, "What are you doing, digging a well?"

He answered, "Oh, no, my man, we don't dig wells in this country."

I asked him how far to Miller House, he said about two or three miles. The old man asked me where I was headed for, and where was I from. I told him, and then bade him goodbye and went on my way. The old man must have said to himself: "What's that greenhorn doing out here? He knows nothing about mining, that's for sure, when he doesn't know a prospect hole from a well!" I met that old fellow later on at Miller House. His name was Bill Cheeseman; he was a real old sourdough.

At the end of my first day out I stopped at a place fourteen miles from Circle. This place was called Reiger's Ranch. Reiger had built quite a few log cabins, barns for his horses and cows, pig pens for his hogs. His roadhouse had three rooms: living quarters, kitchen, and a sleeping room, which was lined with pole bunks. Reiger himself was a large, rawboned man. He looked like a Russian with his foot-long, tea-colored beard. He was a pleasant man, sincere and ambitious. However, he had his ranch located in one of the coldest parts of Alaska, with temperatures registering at times sixty or seventy degrees below zero. Reiger told me that he lost part of his stock the previous winter by their freezing to death, but that it didn't cost much to feed his stock during the summer months.

In the early evening hours Reiger invited me down to the riverbank; he wanted to show me how he fed his hogs right from the river. Birch Creek teemed with fish of all kinds, predominantly pike. As soon as Reiger reached for his fishing pole (he used a spoon hook), it would alert all the pigs and they would come running from all directions. Sometimes they would run between his legs and nearly trip Reiger. It seemed almost as soon as Reiger's spoon hit the water a pike would grab it. Reiger had a club to kill the fish as soon as he dragged it to shore. In the meantime the pigs were all squealing up on the bank waiting for their supper of fish. The first fish Reiger threw up on the bank happened to be a large one—the hogs all got into a fight over it. This same procedure lasted for nearly an hour. Reiger estimated that he hooked out at least a hundred pounds each time he fished for the pigs. After all the fish were consumed, the pigs scattered and busied themselves digging for edible roots; this gave them sort of a balanced diet, I suppose. Small portions of dried king salmon were fed to the cows and horses which had survived the freeze to enrich the native hay he had put up during the latter part of July. Reiger claimed that the dried fish which he bought from the natives was cheap and took the place of grains which were sky-high in price. He never served me any of the cow's milk; he probably knew the milk tasted fishy.

On my second day out from Circle City I made over twenty-five miles. I saw no humans or game. The only weapon I had was a .38 Colt revolver which I had brought from home with me. I ran a good part of the twenty-five miles, arriving at Central Roadhouse long before darkness set in; at that time of year it was getting dark about 7:30 P.M. The Central Roadhouse was operated by a genial old German with a long white beard which made him look like a monk.

My next stop was at a roadhouse and small general store operated by a Mr. Kelly. His supplies were brought in by horse and sled during the spring months from Circle City. This freight was brought to Miller House by Nels Rasmussen, who also owned the winter freighting business; he used horses and sleds with four horses to the sled-load.

My third night I stopped at the Miller Roadhouse, which was owned and operated by a "Cap" Griffith and his wife. The Captain wasn't much help to Mrs. Griffith, as he was a steady customer of John Barleycorn. I believe he would drink anything short of Red Anti-Freeze. Mrs. Griffith had quite a problem keeping her extracts and meat sauces out of the Captain's reach. She never put any meat sauces on the table until she had the boarders all seated. Otherwise, the bottles or their contents would be missing—Worcester Sauce was the Cap's favorite drink when no whiskey was available. He would gulp right down anything that was offered him in a bottle. He died many years ago.

I walked the five or six miles up Mastodon Creek to the place where my brother was cooking. Naturally, we were glad to see one another. Tom turned his cooking job over to me and left for the States to get married. He never did come back to Alaska. Since cooking was my profession, I took over at once. The only thing was that, at that time of year, there wasn't much left to cook. There were shortages of everything in the chow line—it all had to come from Circle City by pack horse at 25 cents per pound.

One day, Nels Rasmussen, who owned the packhorse train, at that time the only one between Circle City and Miller House, packed out to the camp a half-hog from Reiger's farm; you can imagine what kind of pork this would be—fattened on fish and what little grass and roots were available. I cooked a pot roast from the flabby pork; it smelled so fishy while cooking that nobody could eat any. Nels thought he was bringing us out a treat—a black bear would have been much better to eat. Reiger's stock farm turned out to be a failure.

I cooked at the mine until Anderson closed down for the season, about September 20. The miners used wheelbarrows to wheel the pay gravel to the sluice boxes. The miners were paid fifty cents per hour for twelve-hour shifts. I was paid one hundred and seventy dollars per month for sixteen or so hours per day. That same fall when the caribou started their annual migration south it seemed the mountains were alive and moving with them. They just about stampeded the cook shack. The "run" lasted for days.

The cook those days had quite a responsibility. I had to wake up the day crew for breakfast so they could relieve the night crew. The boss, Mr. Anderson, and his wife, had a special breakfast between nine and ten o'clock. I didn't mind this even if it meant extra work for me because they were such nice people. I had twelve men to cook for, with no help. I split the wood, rustled the water, which was handy, besides doing the cooking and baking.

One day, Pete Anderson came into the cook house and told me to fire Fred Schroeder.

I asked, "What are you going to fire old Fred for?"

"Well," Pete said, "he is getting to be too much of an agitator among the men. When he gets up this evening hand him his time, and tell him I am sorry to have to do this before the close of the season. Also hand him this poke of gold dust, it contains twenty-four ounces of dust at sixteen dollars per ounce, which makes $384."

I hated to have to do this to an old friend but orders are orders from the boss. When I handed Fred the poke he had no comments to offer except to say, "Why didn't Pete settle with me himself?"

Pete was that type of man. Schroeder said it was about time he was heading back to his trap line anyway. Schroeder had worked on the night shift.

Nels Rasmussen took a liking to me and offered me a job for the winter. It didn't pay much of course, but it gave me a place to put in the hard cold winter just ahead of us. Nels had about one month's work at a place called "Jump Off," twenty or so miles from Circle City, and a crew of about six men, including Moose Bill, who just puttered around and kept the camp supplied with fresh meat. Nels had a sawmill at Jump Off. We stayed at Jump Off for a month sawing lumber.

On our way from Miller House to Jump Off, we stayed at Central Roadhouse overnight. The man who owned and ran the Central Roadhouse was a fine old German who was called "Old Stead," he was about sixty-five years old at that time.

On the night we camped with Mr. Stead he fried us up a lot of caribou steak for our supper. Being a cook myself I observed how he went about preparing it. While Stead stood over the cook stove watching the steaks of caribou cook he would comb his beard with the same fork he turned the steaks with. He did this unconsciously. I learned a new wrinkle about bread making from Stead. He had a large batch of bread raising in a big tin dishpan with a yellow tomcat sleeping on top of the dough. I thought the cat was sleeping where it shouldn't so I told Stead the cat was sleeping on his bread.

He remarked indifferently, "Oh, that's nothing, there is always a dishcloth between the dough and the cat to keep the dough clean." Stead just glanced over at the sleeping cat and smiled.

I told one of the oldtimers what I had seen. "Oh," he said, "that's where the cat always sleeps."

At supper time this oldtimer, whose name was Bill Hogshead or Moose Bill No. 2, picked one of Stead's biscuits up from the plate and asked, "Hey, Stead, how did all the cat tracks get on your biscuit?"

Stead, the genial old German, just passed if off with a smile.

The old man liked to be kidded. He had a nice garden and raised potatoes, cabbage and carrots. He had a small store in connection with his roadhouse. At mealtimes he always had the table loaded with plenty to eat.

Nels had plenty of groceries at Jump Off to cook with. Four families of natives were camped close to us. They hunted and fished, drying the meat and fish for winter use. We had a funny little old man among the sawmill crew; his name was Noah, and he wouldn't eat anything that had milk in it. I said to myself, "There's going to be a hard guy to cook for." When eating his first meal with us he refused to eat any cake because I made the mistake of telling him the cake contained milk. At each meal he would ask me, "Any milk in this—any milk in that?" I told him I didn't use any milk for cooking anymore for his benefit. His face beamed with pleasure. I put the usual amount of milk in all dishes that required milk; he never complained or showed any bad effects from the milk. I was sure now that the refusal of milk was due to his imagination, and that, undoubtedly, the milk was doing him much good.

One day I made a large custard pie and, of course, Old Noah asked his usual question, "Is there any milk in the custard pie?"

I told him I didn't think so. Noah ate a large section and praised the pie for tasting so good.

But I went a little too far with Noah. Our camp hunter, Bill Hogshead, brought a small black bear into camp. I told Bill I was going to have pot roast for supper, "...and don't mention to anybody about your killing the bear!"

When the crew came in for supper most of the fellows asked, "What's that you're cooking, Jim, that smells so good?"

I told them it was "short ribs of salted pork soaked in water to freshen them."

"We know it didn't come from the Reiger Ranch because we couldn't smell fish."

They just about cleaned up the whole roaster full and said the pork was delicious. After everybody had eaten their fill and shoved their benches back from the table, I asked them if they really knew what kind of meat they ate for supper?

"It was pork, wasn't it?"

I couldn't keep a straight face any longer. I had to burst out laughing, and I told them it was bear. "Bill brought one in late last evening."

Bear or no bear, it was surely good. They asked Bill where he shot the bear and why he had not told them before about killing the bear. I noticed Noah's face—it was white as a clean sheet. Noah jumped up from where he was sitting and barely made the door in his haste to get outdoors. I felt sorry for Old Noah. He was a very sick man and stayed in bed for three days. At one point I thought the old man was going to die. But he eventually recovered and went back to work again, none the worse for his bear supper.

A year-old black bear that has fattened on blueberries is very good eating. It was kind of late for this bear to have fattened on blueberries, but there were still lots of small cranberries to be had. Unlike the "outside" cranberries that are large and grow in bogs, these small cranberries grow anywhere on high ground. In some places the ground is red with them and they are very good eating. The bears substitute the cranberries for blueberries. When Bill dressed out his bear he told me that the contents of the bear's stomach looked like cranberry jam.

I used to try to hunt rabbits in my leisure time, but I was never lucky enough to shoot any. The gang used to kid me for not being able to kill even a rabbit. The next time I went rabbit hunting I slipped half a slab of bacon under my jacket and went off straight to where the Indians were camped. Each morning they brought in many rabbits from their snare lines and they also shot some. I asked them how many rabbits they would give me for the piece of bacon. The Indians were crazy about bacon. They told me to take all the rabbits I wanted from a pile of them lying on the ground. I wanted only the ones which had been shot; snared rabbits wouldn't do—the gang would think I had robbed the native snare line. So I selected six good bloody rabbits from the pile and packed them into camp. They never kidded me any more about my ability as a rabbit hunter. I pot-roasted the six rabbits. They were fine eating for a change. That time of the year, early fall, rabbits are at their best; after the weather sets in, and it gets very cold, the rabbits have nothing to eat except frozen willows; many die during cold spells.

About the middle of November we left Jump Off for

Circle City which was about twenty odd miles from Jump Off. The swamps and lakes were frozen solid enough to support the horses. Nels had four horses and two sleds; he also had a small mule or donkey hitched to a double-ender light sled. Nels used these horses in wintertime to freight supplies out to the various miners around Miller House for the next season's mining operation.

We arrived in Circle City late the same night. It was the time of year when the saloons paid off. All the oldtime miners were heavy consumers of whiskey and beer. I kept house for Nels until spring, as Mrs. Rasmussen was spending the winter in Fairbanks.

A fellow by the name of Clark was making headline news. Clark had just arrived in town from a prospecting trip. He had one dog when he left on his prospecting trip; when he came back he had no dog. He told everybody what a terrible trip he had and how lucky he was to get back alive. He said the thing he regretted most was when he had to kill his dog and eat him to keep himself from starving to death. Three days after Clark arrived in town his old dog, that he supposedly had eaten, staggered into town looking none the worse for the trip; he was, in fact, very much alive.

CHAPTER II
Wintering in Circle City

The oldtimers who wintered in Circle City in the winter of 1910 – 11 were great for "jobbing" one another. One oldtimer named Joe Lagoo, a woodcutter, used to come into town every Saturday night from his wood camp, which was a short distance back of town. Joe owned a very large dog which was as strong as two ordinary dogs, and he had made a small, light sled on which the dog pulled him back and forth from his wood camp. Joe always brought his dog and sled into Henry Power's Saloon with him. As I have already said, there was always a "jobber" around. When they noticed that Old Joe was feeling pretty good and deeply interested in his card game, someone would nail his sled runners to the saloon floor. Lagoo used to have a hard time getting his sled loosened from the floor—he cussed a blue streak at the same time. One time the men rough-locked his sled runners with a dog chain (that is wrapping the chain around his runners). Joe never noticed this until he got home; he just didn't know why his sled was so hard to pull. Joe and his dog were all in by the time he got home.

Another oldtimer by the name of Charley Renny, a little old German, had a hobby of stealing wood from the N. C. Co. wood pile. One day the store manager loaded a block of wood by drilling an auger hole in the block and filling it with black powder. The block was placed conspicuously so Renny couldn't miss the temptation. He packed the block of wood home that same evening. Renny was

31

married to a native woman whose name was Lucy. The next morning when Charley Renny got up to start the morning fire he put the loaded block in his heater stove and crawled back into bed alongside old Lucy to keep warm until the cabin warmed up. It was about 30 degrees below outside the cabin. Renny didn't have long to wait before the explosion. All at once the stove blew apart, landing on top of Charley and Lucy in bed. It's a wonder the cabin didn't burn down; hot coals and ashes covered everything in it. As soon as Renny got his britches on he made a beeline for the commissioner's house. The commissioner's name was Dodson. Meanwhile, Lucy threw snow in the cabin to put the fire out. Charley was all in from running to the law. He told Commissioner Dodson he wanted to have someone arrested for blowing up his stove. Dodson said, "Well, Renny, you tell me who you stole the wood from and we will have him arrested for you. Of course, Charley, you will be arrested also for stealing wood." Incident closed.

Renny had a habit of coming into the old N. C. store each morning to chat with the store manager. On occasion he used to watch Renny whenever he came in; he had a looking glass in the back part of the store and he could see everything that Charley did. Every time Charley passed an open crate of potatoes on the counter he would reach into the crate, grab a potato, and put it into his coat pocket. One particular morning the manager placed a wash basin full of eggs on the counter. As Renny walked up and down the store he passed the eggs and slipped one in his heavy coat pocket, just the same as he did with the potatoes. The manager was watching Charley's every move through his looking glass. After Charley had snatched four eggs the manager thought Renny had enough, so he came walking from behind the counter right up to Renny and says, "Good morning, Charley!" at the same time slapping Charley's two coat pockets hard enough to break the eggs. Charley never let on. He excused himself and said he had to get home quick to shut the heater off (with no eggs for breakfast except the scrambled ones in his coat pocket!). For my first winter in Alaska I had been well initiated in Circle City's society—quite an experience for a fellow eighteen years old.

In the spring of 1911 – 12 I went out to the mines. I had the same job cooking for Anderson as the previous season. I told Anderson I planned to quit about the first of September. The Boss didn't like the idea of having me quit before the close of the season. I told Pete of my big plans of going down to Fort Yukon to trap for the coming season. He only laughed at the idea.

I walked back to Circle City in four easy days. At Circle I ran into an old guy named Black Jack—a "handle name"—his hair and beard were jet black. Ingels was his real name. At that time he was sixty-three years old. He owned some placer ground back of Central on Deadwood Creek, but he never took out much gold. We got quite friendly at Circle City. I told him of my big plans of going down to Fort Yukon to put the winter in trapping. "You better come along with me, Jack," I invited.

He agreed to go at once. I hadn't liked the idea of going alone. I bought a square-stern boat for ten dollars. This style boat was built at Whitehorse, Y. T. There were lots of them at Circle City. People drifting down the Yukon River from Dawson used to abandon them at Circle and walk crosscountry from Circle to Miller House. Most of them were miners bound for Fairbanks. These boats would hold a half-ton easily. The boat was only twelve-feet long with a wide beam and a good flare on the sides, so it wouldn't capsize easily. We bought only food enough to last us for the trip down the Yukon River to Fort Yukon. We expected to buy our trapping outfit there; the distance was eighty miles from Circle.

Rowing the boat, we expected to make the trip down in three days. We carried a pole in the boat, besides the pair of oars. I noticed my partner was hitting the bottle quite often. I was getting afraid he might fall overboard and drown. I suggested to Jack that we better pull for shore and take a nap for a while. I headed the boat for a nice gravel bar. We stayed here for several hours. Jack killed his bottle and sobered up enough to be safe in the boat. Before we left I brewed a pot of coffee for us. Jack felt much better after drinking some of the coffee. Two or more steamboats passed us every day on their way up and down the river. The scenery was pretty much the same all the way. There are hundreds of islands covered with cottonwood, spruce, and willow tree; the river overflows the low-cut banks each spring when the ice goes out, covering many of the islands with six feet of water. All of this area is interwoven with dozens of large and small channels. One has to be a good riverman to follow the main channel. We saw no big game near the river. We passed some fish wheels when we were getting close to Fort Yukon and these were running over with king salmon. There were thousands of geese all along the river. We made Fort Yukon in two long days—we had expected it would take us three days. Unlike Circle City, Fort Yukon's source of income was derived solely from dry fish and furs. Fur-bearing animals of all species were plentiful. Fur wasn't high those days and neither was grub. You could buy an outfit containing anything you wanted for about one-third of what it would cost you these days. When we arrived at Fort Yukon a small crowd of oldtimers, white men and Indians—mostly old Indians and some children—were on the bank. The women held back for a while so they were the last ones who came out to the river bank to size us up with the rest of the population. We tied our boat to the shore and climbed up the bank. They all continued staring at us with curious eyes. They thought at first that we were just a couple of floaters drifting down river. They started asking us all kinds of questions. They wanted to know where we were going and where we had come from.

There were a few log cabins, mostly old. Some were close to the river bank and some were built farther back from the river. Most of the cabins seemed to be unoccupied. The owners were probably out fishing. Most of the natives here were living in tents along the river. Two families generally lived at each fish wheel. They would catch enough king salmon in three hours to keep them cutting fish for the rest of the day. Today, they can't get enough king salmon to eat due to too much commercial activity. Years ago, at the mouth of the Yukon the fishing was so heavy that it has resulted in the present short runs of the king salmon.

The main street in Fort Yukon, First Avenue. (Courtesy Carroll family)

We told everybody that we were going up the Salmon River to trap. They all told us that there was nothing up that river except brown and grizzly bears. We told them bears or no bears we were going up the Salmon River. We bought another square-stern boat. By butting the flat sterns together we had a boat twenty-four feet long. The people all told us that we would never make it up the Salmon River with that kind of a boat. Even this didn't discourage us, any more than the bears did. The Salmon

River flows into the Porcupine River forty miles northeast from Fort Yukon. The Salmon River flows from the north, heading from the Endicott Range and is about three hundred miles long.

We bought our trapping outfit from a trader named Harry Bean Cook who was very cooperative. We didn't waste much time at Fort Yukon. It was late now for us to be going so far, but we intended to keep on going until ice formed in the river and stopped us. Our only motor power was two poles and a pair of oars and two hundred feet of quarter-inch line (this was used to pull the boats from the shore; one man stayed in the boat to keep the bow away from shore). We bought two big dogs at Fort Yukon; the dogs for the most part ran along the shore. Sometimes we had them in the boat. We were soaking wet to the waist every evening from wading in the icy waters. At nearly every riffle we came to we had to jump into the ice-cold water up to our knees and boost the boat by hand over the swift shallow places in the river.

Our first day out from Fort Yukon we made about five miles up the Porcupine River. We kept going until we were forced to stop on account of ice starting to form in the river. We had made eighty-six miles upstream, forty-six miles up the Salmon River and forty miles up the Porcupine. We stopped at a place where there was plenty of dry wood and logs to build a cabin. The day after landing I went moose and bear hunting. We saw no tracks but I went hunting anyway, thinking I might run into something. I was gone from camp possibly two hours—when I heard brush cracking. I thought that it must be a bear or a moose. I crept up carefully from the direction the sound was coming from, and who did I run into but Old Black limbing a house log.

I asked, "What in hell are you doing up here, Jack? I almost shot you for a bear."

"Well," he said to me, "why did you come back so soon?"

I said, "I don't know, Jack, I never left the river bank. I must have traveled in a circle." I was completely turned around and lost. I asked Jack where our camp was.

"Right over there." He pointed a finger at it. This was how we discovered that our cabin was being built on an island! We got moved into the cabin November 1st. Its size was twelve by fourteen feet; it had a flat roof, pole floor, hewed-out door, two small windows, twelve-by-twelve inch window panes. One window on each side of the cabin lit up the interior quite well. During the shortest days we had to burn a candle all day to be able to see anything. The cabin was very comfortable compared with roughing it out in a flimsy tent.

It was now about time for us to commence looking around for some fur signs. We had two large dogs and one small toboggan, which was sufficient to haul our bedding, traps, and grub. I had done a little trapping before on my father's farm back in Minnesota. Most of the old miners, where I had cooked, did a little trapping during the winter months. They had showed me how to set traps for the various fur bearers and what to use for baits. One was supposed to have a different bait for each animal. The marten is the easiest to catch. Next in order would be the mink and lynx; the lynx is the dumbest fur animal.

I remember one time I caught a lynx in a snare. He had twisted himself loose from the snare, which was made from eight-ply picture wire. After getting loose from the snare, he didn't have sense enough to run away; he was lying about eight feet from the chewed-off snare. I looked at him for a couple minutes, figuring what to do. I had a big rifle with me, but I didn't want to shoot him with it, as it would damage the skin too badly; so I cut a good willow club and swung at him. He jumped at me the same second. The lynx and my club collided in mid-air. The blow killed the lynx. I found out the best way to kill a lynx and not get the fur bloody is to have an extra snare in one's pocket when you have a live lynx in either a snare or a trap. Cut a willow pole about eight feet long and tie your snare at one end of this pole; walk up to the lynx carefully and slip the snare loop over his head and pull on the snare; he will be choked in a couple of minutes or less.

Black Jack had done a little trapping near his mining claim at Deadwood, never catching much, of course. I used to let Jack trap from the cabin, running short lines in several directions. In this way he could be home in the cabin every night. Jack was an old man while I was only an eighteen-year-old kid, and trapping was a picnic for me.

I decided to do my trapping up river from our cabin. The first day up river I had the two dogs loaded with traps, snares, bedding, grub and dog feed. It made a top-heavy load because the toboggan was too small. I took no tent or stove along, but camped out in the open by a campfire. The first day up river I saw quite a few mink and lynx signs. I didn't expect to see any marten signs. I knew they were to be found back from the river on the timbered draws, ridges, and slopes.

What I didn't like about camping out in the open was getting up in the morning with the temperature possibly twenty to thirty degrees below zero and the bedding white with frost. When camping out like this I always had some shavings and kindling wood handy for morning. Instead of crawling out of my caribou bag to fire up, I moved with my bag over to where the fire was; when the fire got hot I crawled out of my bag and dressed for the day. I always slept on top of my clothes; this helps to keep one warm. The first thing I did after getting up was to melt snow for water to brew coffee. After that, I made one pan-size hot cake, and chipped some butter loose to put on the hot cake while it was still warm; I then fried about four chunks of bear, put some sugar on the hot cake and ate breakfast. With those temperatures one has to eat fast before one's breakfast freezes up on him.

CHAPTER II — Wintering in Circle City

After four days up river I returned home and found Jack in fine spirits, he had walked out in four directions and had found quite a few marten signs in each direction he went. I brought in with me one lynx and two mink and I made quite a few sets on my way down river. I had gone up twenty-five miles. The lynx I caught looked very big to me; it was the first one I had ever seen. I was slightly scared of it; scared of a lynx that was too dumb to run away!

A few brown and grizzly bear signs began to show up. They were apparently tearing up the shore ice for dead salmon that had floated from the spawning grounds farther up the river. The first track I saw in the snow I thought had been made by a man on snowshoes; but when I got close to the track I saw by the claws it was made by a bear. It was late for these beasts to be out. They broke up the shell ice that was four inches thick in places. This recalled to mind what the oldtimers had said down at Fort Yukon about the Salmon River being a bear paradise. Evidence of what they had told us was surely showing up everywhere. I thought to myself that almost all bears should be in hibernation by that time. An old Indian story says bears that stay out all winter fishing in open water are dangerous; they get themselves coated with an armor of ice which a bullet cannot penetrate. I don't know how true this was; anyway, I didn't like the idea of camping out with so many bear signs around. Of course, I had the two dogs with me but they were cowards—more scared of the bears than I was.

I started up on my second trip after being gone six days. There was lots of lynx sign all along the upper river. On my way back home I picked up four lynx and six mink. Jack had better luck than I did. We were both happy to be home together. Jack was excited; he told me that he had discovered a new way to catch lynx. He said we should keep it a secret.

I asked Jack how it worked, "Do you put salt on the lynx's tail?"

"Oh, no," Jack said. "You just wait until I explain it to you. Yesterday, I was about two miles from home when I discovered that I had our looking glass in my pocket and I didn't want to pack the glass all day. So I made a lynx set with brush and placed the glass in the back of the lynx's house, using the looking glass for bait. Would you believe it, on my way back, a matter of about three hours, a lynx had come up to his set and danced all around it thinking it was another lynx he was seeing?"

I had to laugh at Jack, telling him to leave a comb at the set next time he visited it so the lynx can comb his whiskers.

A bustling street scene on First Avenue in Fort Yukon, circa 1912. (Rivenburg Collection, UAF-1994-70-279, Archives, University of Alaska Fairbanks)

The only thing we had in the meat line was rabbit. We had about two hundred dollars' worth of furs between us, which wasn't bad. It took quite a few skins those days to bring two hundred dollars. The prices were: lynx $4 to $5; mink $2.50; marten $4 to $6; and cross fox $7. At this time Jack and I decided to make a trip to Fort Yukon. It would take us about ten days to make the round trip. We were running short of certain things in the "eats" line and wanted to replenish these shortages. We were just about out of candles; lately we used candles

CHAPTER II   Wintering in Circle City

only to cook and eat by. By leaving the stove door open we could see well enough to skin animals. We always cut the mink's scent bag or the strong musk odor would drive us out of our tiny cabin or give us a headache.

At Fort Yukon we expected to buy two more dogs. This would give us dog-power enough to ride once in a while. When we arrived in town everybody was out with a glad hand to greet us and tell us how lucky we were that the bears hadn't got us. Jack got pretty drunk our first night in town and, of course, he had to tell everybody about his new lynx set. He wound up by buying all the looking glasses that Harry Beancook had in stock. After that he was dubbed "Looking Glass Jack."

It was getting close to Christmas and the thermometer was hovering about forty to fifty degrees below zero.

First Avenue in Fort Yukon. Large first building on right is the theater/show house, second building on right (with porch) later became the territorial government schoolhouse. (Courtesy Candy Waugaman)

We made new plans to stay in town until after the holidays. A few old trappers, who trapped close to town, used to come into town also, sell their meager fur catch, and buy a little grub. They would have a jingle in their pockets to buy one or two bottles of whiskey and, possibly, play a little poker in the back of Beancook's store. It was a regular hangout for all the oldtimers. Those who didn't play poker would spin yarns or sing unprintable songs. There were at least several white men there married to native women.

New Year's at Fort Yukon was always the Chief's day. Chief Robert was a handsome native then about forty-five years of age. Incidentally Chief Robert passed away this last fall at the age of one hundred. He could remember when the Hudson Bay Co. was at Fort Yukon. Chief Robert used to give a potlatch for everybody. This big feast used to last three or four days with lots of eats for everybody: moose meat, fish, caribou, ducks, geese and even some porcupine. This was all gathered in the late fall by the Chief and put in his cache, not to be touched until New Year's. There was dancing every night. They danced to oldtime violin music. There were always two fiddlers at a time playing together their oldtime dances. The Chief was quite a playboy. Everybody had to dance. The Chief was always his own floor manager. He wanted everybody to have lots of fun. He was such a good-looking chief that he was very popular with the womenfolk.

At about this time a commissioner for Fort Yukon was appointed. There was no marshal there then. Whiskey was forbidden among the natives. A whiskey agent for the Indians used to pass through Fort Yukon once in a while, stay a few days, and get drunk as a boiled owl. It was said that he used to bootleg to certain Indians—the ones he thought he could trust. I remember an old native was arrested and dragged up before the recently appointed commissioner and charged with being drunk and disorderly. He was advised that if he told the commissioner who got him drunk he would go free; otherwise, he would have to go to jail. Big Felix was the native's name. He was a big, tall Indian. He told the Commissioner how he had stayed up late that evening visiting friends.

"It was dark, I have hard time to find trail. I fall down—lots of time—every big snowdrift. Last time I fall snowdrift my hand feel something hard and smooth—I grab hold of it; it was bottle. I loosen cork; it smell like whiskey; I remember long time ago I smell whiskey—this one just same smell. I don't know what to do. I think better I take big drink now. I think better I go visit more my friends. Before I go in friend's house I take one more big drink. After that I don't know what happen. I wake up in small room, iron on window. Someone stay with me. That is all I know—true to God."

The Commissioner said, "Well, Felix, I know everything you have told me is a lie, but I guess I will have to let you go this time for lack of evidence." Poor old Jim Felix died of the flu in 1926.

There was another old gentleman there named Bomont, who had been there since the Hudson Bay days. He died more than forty years ago. He had a little store at Fort Yukon. If you came to his store with a purchase from another store he wouldn't sell you anything—he was funny that way. He was a fiend to play poker. He used to have a game every night in his living room, midway between his store and storage room. One evening the oldtimers who played almost every night with Bomont asked me to go along with them to "Dad's" place. He was commonly called "Dad"; he was generous with his whiskey, too. Dad Bomont was a small man, a French Canadian. His hair was snow-white. He must have been seventy years old. He had one fifty-gallon oak barrel full of whiskey in his back room with a modern faucet driven in the end of the barrel. He never sold any of this whiskey to my knowledge. One night when the poker game was going full blast somebody slipped into his room with a bucket, intending to draw off a gallon or so. About the time the whiskey thief had his bucket almost full, he thought he heard someone walking towards the door of the whiskey room. He got scared and beat it out the back door. In his haste to get away he forgot to turn the faucet off. It wasn't long after the thief made his getaway until a small wave of whiskey came creeping under the doorsill into the room where the poker game was still going on. Dad jumped up from the card game and rushed into the whiskey room and turned off the faucet. There must have been two gallons of whiskey on the floor. Dad was sure some of his friends had pulled the trick. He cussed everybody out, calling them all a bunch of thieves and crooks. This all happened before the Volstead Act.

On another night I accompanied the same gang to Dad's store for more poker-playing. They took advantage of old Dad this night. Dad had a habit of closing his eyes tight every few seconds. Some of the players watched their chance and when Dad closed his eyes they used to slip one another cards across the table. Dad lost quite a bit this night. He had to draw on his safe for some more money. Dad came from his safe carrying a canvas money bag, containing almost $300 in silver. Dad emptied the sack of silver on top of his big round poker table, but before he had time to pick out the silver he wanted, a big, good-natured fellow by the name of Abner kicked the table over spilling all the silver on the floor. The silver scattered everywhere—under tables, stoves, beds, and through cracks in the floor. Dad never did forgive Abner. Of course, everybody tried to help Dad pick up his silver, but the only silver Dad saved was what he picked up himself; the rest pocketed what they picked up. Dad was outraged; he called them all a bunch of robbers and horse-thieves.

"You accepted my hospitality only to have a chance to rob me. What kind of a bunch of outcasts are you? I want you all to get out of my place and stay out!" Less than a week later Dad had the whole bunch back again; that is, everyone except Abner.

Bomont used to make an annual trip out to the States once a year. Every time Dad returned from the States he brought a good-looking girl with him. He used to tell her he was a rich trader from the far north and that he needed a girl to act as his secretary. He told her also that he was getting old and couldn't attend to everything himself; he further told her he would will all his property to her, and when he died she could simply take over. The first girl he brought in with him he kept indoors all the time. On Dad's next trip to the States he left his farm girl here to look after his business; but before Dad got back again this girl had left and married someone. On Dad's next trip back he brought in another girl with him, feeding her the same line of promises he did the first one. Dad worked it pretty good—he never had more than one girl on hand at a time.

The second girl he brought in never did leave the house. It looked as if he didn't want her to meet any of the roughnecks for fear they might put bad ideas in her head. On Dad's third trip to the States the second girl he brought in married a trader during his absence. This girl didn't do so bad—she had everything Dad left in the store moved into her new husband's store. And, of course, they sold it as their own merchandise.

The third girl came into Alaska alone on a boat down the Yukon River via Skagway. She was advised that old Dad was in trouble and had been arrested for white slavery. A purse was collected from the passengers and officers of the boat; it amounted to enough to take her back home. She had to wait for an upriver boat—there were no airplanes at that early date, no roads or railroads; some people walked from Fairbanks to Valdez or Saint Michael and boarded a boat for Seattle. I never learned the name of this girl. A missionary lady, Miss Woods, took this girl into her small mission and looked after her until a boat arrived for up river and home. After two years elapsed old Dad arrived back in Fort Yukon—broke. He came into Beancook's store one day and shook hands with everyone except Abner. Dad refused to shake his hand.

Beancook spoke up and said to Dad, "You know Abner don't you, Dad?"

"Yes," he answered, "I know him all right; I know him too damn well!"

Dad sold his property. He was offered very little for it, just about enough to take him back outside. We never heard from him since then—nearly forty years ago.

CHAPTER III
Back on the Trap Line

It was now January 1, 1912. I had a hard time getting my partner Jack started back to our trap line. Someone always played jokes on Jack every time he got drunk, which was often. They used to paint his face with different colors of paint while he was laid out drunk. Some-

1925 Fort Yukon. At center left is Horton & Moore's store, one of four trading posts in Fort Yukon at the time. With Hospital, Mission House, and Church in background. (Courtesy Carroll family)

times he wouldn't notice his face unless he happened to pass a looking glass on the wall. He came into the dance hall one night with his face all painted—he nearly scared the wits out of all the children. They had put stove blacking around his eyes. He used to do a lot of cussing, which did him no good. I must admit I was in on the painting job myself. We used to tell Jack not to get drunk again and nobody would paint him any more. At times this paint was hard to get off his face; we had to use turpentine to remove it. It was getting to be no joke with Jack. He threatened to kill the whole bunch of us if we didn't stop painting him.

Jack finally did sober up and we bought two more big dogs and a larger toboggan. We picked up what shortages we had in the grub line. We were going back in style—Jack could even ride when the trail was good. I used to exercise the dogs every day in Fort Yukon to keep them hardened in; that is, I used to drive them around town for an hour: this kept them in trim for hitting the trail. We made it up to our camp in four days. The weather wasn't too cold and the trail was fair; it hadn't snowed much during the period we were away. After we got home Jack used to say, "No more Fort Yukon for me." Jack was an old man between sixty and sixty-five years, while I was only eighteen years old—just in my prime.

Jack asked me if I wanted to take some of his looking glasses along with me and make some of his looking-glass lynx sets.

I said, "No, Jack, I would rather not bother with any."

After a couple of days' rest we started out on our respective lines. I picked up a skin or two here and there. Most of the animals caught were badly ruined by mice and ravens. It looked as if the brown bears had all gone to bed. There were no more fresh signs. I went about twenty miles further upstream; there was nothing to it with four big dogs. Jack's lines were all short and fanned out from the cabin like spokes in a wagon wheel—each of his lines were short enough to allow him to make a round trip each day. It was eight days before I got back from up river. Jack had commenced to get a little worried about me for being gone so long. He thought a

CHAPTER III — Back on the Trap Line

brown bear might have caught up with me, or that I had fallen in the river and drowned. I picked up quite a few furs up river. Some were badly damaged and needed sewing. Jack did all right on his short lines. Some of his marten also were damaged by mice and squirrels. We spent a few days at home, sewing damaged furs and stretching the undamaged ones. If the trader we sold the furs to didn't notice the damaged skins it wasn't up to us to point them out; we were selling furs, it was up to the buyer to examine what he was buying.

I remember one time I had three poor and damaged cross-fox. One had no tail, another had no head, the third one had badly rubbed hips. I told Jack that I was going to try to make a whole cross-fox from these three damaged ones. Jack didn't think much of the idea. One of the fox had a good head and shoulders, one had good hips, and the third one had a good tail and hips. I trimmed off all the ragged edges and rubbed places. I was good at sewing, considering the short experience I had had the past winter. I simply sewed the head and shoulders together and then I fastened on the good hips and tail and pulled it loosely over a stretcher. The whole fox pelt was probably a little shorter than average length.

I looked at it and said, "It will be adorning some lady's shoulders someday."

When we were showing our collection of furs to a fur buyer he picked up this tailor made cross fox first, shook it out and said, "Now that's what I call a nice little cross!"

I was afraid if he shook it much more the tail might drop off. Cross foxes are easy to match.

While up the river on my last trip of the trapping season, I picked up all our traps and snares and I hauled everything down to the cabin. The snow was getting wet in midday. We pulled our boat up on top of the bank to be safe from high water and ice. What little grub we had left we put in cans and hung up in spruce trees with wire. Wolverines and bears couldn't bother it then.

On my annual moose hunt each fall up the Salmon River I stop and have a look at the old cabin that Jack and I built forty-five years ago. The cabin is still in good shape.

That year we left about the middle of April for Fort Yukon with a big top-heavy load consisting of our bedding, furs, camp equipment, etc. We each had a caribou sleeping bag with hair on the inside. These were very warm. One could safely sleep out in the open with temperatures forty or so below zero. We had a small piece of canvas to cover the bags after we crawled in for the night. We took four easy days to come into Fort Yukon.

This photo of Haly's Roadhouse was taken in about 1905. The generous Jim Haly owned and operated this Fort Yukon roadhouse from 1901 to 1918. (University of Washington Libraries, Special Collections, UW 24104z)

Everybody asked Jack how his looking glasses worked as lynx bait and how many lynx he had caught with them. We intended to stop for a while at the Haly Roadhouse. Jim Haly was a kind old French Canadian. He had married a native woman many years before. He came to Fort Yukon in 1901 and operated the same roadhouse until 1918. Jim never turned anybody down for a meal or a bunk to sleep on. If you had no money you could

CHAPTER III — Back on the Trap Line

stay at the Haly House as long as you wanted to. This generosity kept him more or less broke all the time. Jim's credit was always good at the local stores and he always managed to pay his bills. Jim Haly and his wife had come into Alaska via the McKenzie River over the Rat River portage, then down the Porcupine River to Fort Yukon.

After staying a few days with Jim we rented a small shack for $5.00 per month. After moving into the little log cabin we offered our furs for sale to the highest bidder. We had the buyers come to our cabin to examine the furs. A trader by the name of McElroy offered us the highest price, so we sold our furs to him. He paid us a little over four hundred dollars. Pretty good for two cheechakos!

Old Jack and I dissolved our partnership at this time. He said he was getting too old to rough it any longer in the sticks. He bought a horse from one of the steamboat woodcutters and went into the wood business for himself in Fort Yukon. He used to charge five dollars for a load of stumps and windfall. He cleaned up around three thousand dollars and left for parts unknown. I never heard from him so I was on my own again.

A fellow by the name of Joe Ward and I got to be good friends. Joe was an Old Country Englishman. In fact, over three-fourths of the white men in Alaska at that time were Scotch, Irish, or English. The whites, most of them, came into the country via the McKenzie River route. Joe Ward had a partner named Fred Schroeder, a German from the old country. Joe's trap line was located ninety miles up the Porcupine River. (Joe has trapped in the same place since 1909 very successfully; he still lives on the same trap line today.)

Joe and Fred invited me to go up to their place and "spring out" with them during the ice-break time, which is the time of year the ice breaks up in the Alaskan rivers, usually around the first of June. The ice flows downstream causing ice jams and floods on its way to the sea. I went up with Joe and Fred. They told me not to take along anything in the grub line and they would give me dried salmon for my dogs to eat. I just had to throw my caribou bag in the toboggan with a change of clothes. I rode the ninety miles behind their dog teams. We made the trip from Fort Yukon to Joe's place in three days. At Joe's and Fred's camp there was nothing to do but wait for the ice to go out. We used to shoot all the ducks and geese we could eat, almost without leaving the door. I don't know how Joe Ward got teamed up with Schroeder.

(Standing left to right) Harry Horton, Waldo Curtis, Bill O'Brien, Rube Mason, John McNichols, "MP," Bill Mason, James A. Carroll, Fred Schroeder. (Seated left to right) Pete Nelson, Joe Ward, Jacob Thomas (known as Tommy the Mate), the little boy is Jacob's son Billy Thomas. Taken in front of Harry Horton's store. (Courtesy Carroll family)

Of course, Joe didn't know much about trapping in those years and Fred knew *nothing* about it. But Fred promised Joe he would show him how to trap all animals—that he, Fred, knew all there was to know about setting traps and woodcraft and he could also tell by an animal track whether it was made by a lynx or a marten. Joe had no reason to doubt what Fred told him.

One day the previous winter, Joe told me, he was scouting around for anything he might run across when he came

CHAPTER III

Back on the Trap Line

to a round track made in the snow. Joe didn't know what made the track so he hurried back to tell Fred about the large, strange tracks he had seen. So Joe got Fred to come back with him to tell him what animal made such a track.

After Fred sized up the track for a while he turned to Joe and told him the track was made by a mountain lion. "If we can catch him his long tail will make a fine stew." (Incidentally there are no mountain lions in Alaska.) Joe says that at the time he believed what Fred said. Fred used to take a trip every day to look at the trap, a No. 4 one, and the two snares.

Fred did come back one day with a big lynx slung across his shoulder—happy as a child. He was so tickled by his catch that he hollered to Joe before he got to the cabin that he had caught the cougar, or mountain lion.

Joe came running out of the cabin and down the trail to get a quick look at the lion and asked, "Where is his long stew tail?"

The mountain lion turned out to be a lynx with a tail two inches long. Fred had no comments to make.

During their first year of trapping, Joe told me, after a light snow there were hundreds of rabbits everywhere, even down on the river where they had a hole chopped through the river ice for a water hole. Neither of them knew what made all the tracks in such a short time, during just one night. Fred called them caribou tracks. They both started hunting at once for the caribou. They would hunt all day and not see even a rabbit. The next morning there were ten times more tracks. The whole area was just padded down. They had built their cabin that fall. The rabbits used to come at night to feed from all the green slashings, such as green spruce tops and willows, which rabbits seem to go crazy over. Joe says they both went hunting again without seeing a sign of caribou and that he couldn't understand how there were so many tracks but no caribou sighted. Joe also said he came across a beaten-down "caribou" trail where it ran under a log about six inches above the ground. Schroeder's caribou trails all ran under this low log.

Joe turned to Fred and asked him, "How big are these damn caribou anyway?"

Fred answered, "Oh, I would say they stand five feet tall, more or less." That was all Joe wanted to hear. Fred never mentioned caribou either after that.

On June 6th we started floating down the Porcupine River, ninety miles to Fort Yukon. The only incident that occurred was that a brown bear tried to get up a steep cut bank but couldn't make it. We didn't want the bear.

Fred Schroeder is at far right. Man holding the child is Fannie's father, Dick Martin. Man standing at farthest left is Old Rodrick, Dick Martin's father-in-law. Woman on far left is May Martin, Fannie Carroll's mother. Middle lady is Mary Roberts. (Courtesy Carroll family)

Fred took one shot with his rifle and missed. The bear then started for the boat. Luckily, we had a heavy scow-shaped boat. The bear tried to climb into the boat. We shot the bear with a shotgun. He sank like a rock in the swift muddy waters of the Porcupine. Fred got in only the one shot, as his rifle jammed. There was lots of excitement for a few minutes in handling the boat; we only had one pair of oars and a pole.

CHAPTER III

Back on the Trap Line

After we arrived in town I stayed at Jim Haly's Roadhouse for a month. Old Jim was a great soup maker. He had a large seven-gallon soup kettle that he always kept full of soup. I often wondered afterwards how he kept so much soup on hand without it souring on him. He must have kept adding soda to it. He used to put all kinds of small game in his soup, such as squirrels, ducks, ground squirrels, rabbits, and sometimes a piece of moose or caribou meat that the Indians happened to bring in. They were all good to Jim in this respect. Jim always had soup to serve any time of the day or night.

One time Jim had his soup pot and slop bucket sitting side by side. By mistake a fellow by the name of White emptied his wash basin in Jim's soup pot instead of the slop bucket. Apparently, his soup was popular as ever. At times Jim had nothing but soup to serve. Haly's soup was known from Dawson to Saint Michael. Word about it was carried along by the steamboat crews.

I decided to go trapping alone the next season up the Salmon River. Beancook gave me an outfit of grub too on credit to go trapping with. Otherwise, I was broke. Jim invited seven of the oldtimers up for supper one evening, as he happened to be feeling good that particular night. All the guests were feeling good. It could have been Jim's birthday. He had the table loaded down with all kinds of good eats: moose roast, duck and potatoes, bread and applesauce.

Sourdough Bill was also there. Bill was quite a character—a native, and always hungry. Bill had been crippled in his early youth and had a difficult time navigating. He was always around Jim's at mealtime; we all told Bill to get under the table this feast night and we would pass him all the food he could eat—out of Jim's hearing of course.

So Bill crawled under the big round table. Jim couldn't see Bill under the table, the cloth just about reached the floor all around. So the boys passed almost all the grub on the table except the dishes under the table to Bill. Old Jim barged into the dining room from the kitchen.

Seeing the table bare of grub he remarked: "By golly, you fellows must have been hungry. Well, youse have some more of everything; there's lots of soup left in the kitchen."

Meanwhile Bill was having the feast of his life. We all got up to go and asked Bill how he was making out.

Bill said, "Fine. All I want is some more tea." (Bill used to drink about a gallon of tea every time he ate a meal.)

During the winter months Jim Haly used to buy rabbits by the hundreds. He had rabbits stacked up like cord wood in his cache. He never took a chance of running out of soup stock for his famous soup. As mentioned before, Jim was married to a native woman. She never bothered with any cooking except for making certain dishes for herself. A year or so later Jim had some competition in the roadhouse business. A man named Shuman, a former cook on one of the Hudson Bay boats, came to Alaska via Fort McFerson over the Rat River portage, and down the Porcupine River to Fort Yukon. He built himself a roadhouse. He was an excellent wildgame cook. Shuman was a fat man weighing two hundred and fifty pounds, and shaped like a big barrel. He always looked untidy in his dress and he was anything but clean in his cooking.

I remember Curley Well and I stopped at Shuman's overnight. We told him we would like to get an early start in the morning. Shuman set his alarm clock to ring at four o'clock. When the alarm went off it woke me up too. So I just lay awake in bed until breakfast was ready. Shuman got right up and made a hot fire in a big cast-iron, flat-top heater he had. I watched Shuman all the time; he didn't know it. First thing he did toward cooking our breakfast was to spit a mouthful of tobacco juice on top of his cast-iron heater. After this burned out dry, he made our toast on the same stove. As soon as Curley got up I told him of the incident. Shuman wanted to know why we were not hungry that morning. We told him we had drunk some bad whiskey the evening before, and that we would settle for a cup of the coffee that smelled so good.

Curley used to trap below me on Salmon River. Jim Haly was caught in the dragnet of 1912 – 13, when Fort Yukon moved into Fairbanks temporarily. When Jim got back to Fort Yukon he told about all the roadhouses he had stopped at on his long trip to Fairbanks and back. He remarked that not one of the roadhouses compared with the Haly House. He missed especially the soup. His old Mary carried on the duties of the roadhouse. Jim's Roadhouse had a dining room and lobby in the front section; in the middle section was the kitchen; and the back part was the bunk house, lined with tiers of pole beds on each side. His mattresses had large holes in them, some of them eight inches in diameter, caused by having been through so many floods. The tourists' boats used to tie up for a while right in front of the Haly House and Jim took pride in conducting the tourists through his roadhouse. When they saw those large holes in his old mattresses, not even covered over with any kind of a blanket, one could hear the tourists remark: "My, oh, my, how could anybody sleep in such beds?"

At that, Jim was a kind old soul and always meant well. The tourists were always advised by some joker not to forget to be conducted through the Haly House. "It is one of the main attractions at Fort Yukon."

Some of the tourists used to ask foolish questions, such as, how far it was back to the ice wall? We had a character in town (every small place has one). His name was Tony; he hailed from the Azores. Tony was always there when a boat arrived. He mingled with the crowds. Every time any tourist asked a tough question, he was always referred to Tony.

"See that man over there with the black mustache? He

can answer all your questions." Tony's complexion was very dark. One old tourist lady even asked if he belonged to this native tribe. The tourists used to ask Tony also what he did for a living. Tony answered them by saying he was a trapper by trade, and that he cut some steamboat wood, and ran off a batch of "White Mule" once in a while. The tourists did not know what "White Mule" was, so Tony had to explain to them.

One time they had Tony stuck for an answer when they asked him, since he was a professional trapper, how many times you could skin a fox in one season. Another tourist

On right is the Fort Yukon government school. At left is the teacher's residence. (Dr. Cook Collection, UAF-2003-109-12, Archives, University of Alaska Fairbanks)

would ask Tony how he kept his cabin warm during the 70 degrees below zero weather. Tony told them when it got to 70 below zero his stove wouldn't draw, so he had to keep a man on his roof shoveling the smoke which solidified as soon as it hit the 70 below temperature. Another would ask Tony if the Yukon River always ran downstream. Tony told them he had heard of the Yukon River backing up many years ago. A lot of these questions had poor Tony tongue-tied. He didn't want to embarrass the tourists too much by answering all these questions. He had to say "I don't know," once in a while.

Once a tourist, pointing his finger at H. Beancook's store, asked, "What do they do in that building?"

Tony answered, "Oh, that's just a skinning house." The tourist wanted to know what a skinning house was. Tony told him that a skinning house was a store where the trader skins the trappers.

There was a Miss Wood in Fort Yukon whom I should have mentioned before. She was a kindly old missionary lady. She came to Fort Yukon about 1905 and ran a small

Fort Yukon Post Office. (Courtesy Candy Waugaman)

private mission that I knew of or saw. She had a small church and school combined. She taught many of the older Indians how to read and write. Lots of clothing and blankets were donated to her from "outside" sources for distribution to the needy natives, children and grownups. But no one was privileged to receive more than the next person. Miss Wood was also a registered nurse and gave medical aid to those who needed it. She always had two or three girls staying with her. These young folks she raised in her mission. She was postmaster for many years. During the winter months our mail was hauled by dog team via Fairbanks to Circle City, then on down the Yukon River to Fort Yukon. These mail carriers were a hardy bunch of men. They kept the mail moving regardless of weather conditions: extreme cold, blizzards and snow. They never failed us.

Ed Biedeman, a pioneer mail carrier, carried for over twenty years. He got into an overflow one night between Eagle, Alaska, and Circle City and froze both feet. The weather was down to forty degrees below that night. All of his toes on both feet had to be amputated. After many months spent in a hospital Ed was back hauling mail as usual. The Government wanted Ed to retire on a pension. Ed refused and carried on with the mail despite his handicap. Ed was stubborn and carried on until he got too old for the trips. Ed has been dead quite a number of years now; his family still lives in Eagle.

Our mail in the summertime was brought to Fort Yukon by steamboats via Skagway over the White Pass route, then down the Yukon River. One day, I was in the post office when a mailman named Anthony came in with a pouch of mail slung across his shoulders. He had just floated down from Circle City with the first-class mail. Miss Woods asked Anthony how he came down.

"I floated down from Circle, Miss Wood."

She asked, "Do you plan on floating back again?"

In those days we had no motive power except a pair of oars and a pair of strong arms.

About this time a man named Frank White was brought in from the Chandalar country suffering with scurvy. Miss Wood took White into her small mission house and nursed him back to health again, after which she married him. White was about half Miss Wood's age. He built her a nice two-story Mission. They had just about moved into their new building when it caught fire from the basement and burned to the ground—they lost everything they possessed. But this didn't discourage White. He went right to work and built his wife another mission, all from logs, of course. Everybody donated something to them, mostly clothing. Shortly after the new mission was finished Mrs. White died—in 1916. White then sold the place, thus ending the story of a noble soul whom everybody loved. Mrs. White was buried in the old Hudson Bay graveyard at Fort Yukon, Alaska. Shortly after Mrs. White died, Frank White went out to the States and never came back again.

A Fort Yukon woman walks near Miss Woods' two-story Mission House. St. Stephen's church can be seen at left. (Walter & Lilian Phillips Collection, UAF-1985-72-34, Archives, University of Alaska Fairbanks)

CHAPTER IV
Pulling Out

In August, 1912 and 1913, I decided to pull out for the trap line. The outfit of grub Beancook advanced me amounted to $130. I never took any bacon or lard with me. I figured on using bear for bacon and lard. The boat I had picked up would hold about six hundred pounds, not counting the dogs, when I left Fort Yukon. The dogs ran along the shoreline. It took me six days to reach my cabin—the one Black Jack and I had built. Nothing we had left in the cabin or tree cache had been touched. It was the same as when we had left the previous spring. Pulling and poling a loaded boat upstream is hard work. I thought up a bright idea—why not have the dogs pull the boat upstream? I tied two quarter-inch lines to the boat; one I tied three feet back from the bow, the other line I fastened to the stern of the boat. By pulling on either rope I could steer the boat well. If I wanted the boat to head out I would pull the stern line; for the boat to come closer to shore I would pull on the bow line. With the dogs pulling on their line we would go right along. I ran along the shore and kept the lines from tangling up with snags and brush. There was nothing to it now. Sometimes it was all I could do to keep up with the two dogs. When I came to the head of a sand bar I would holler to the dogs to get in the boat, then I would pole the boat across to the next bar. As soon as we hit the opposite bar the dogs would hop out and have the line stretched out before I had time to get out of the boat myself. The dogs learned quickly what I wanted them to do. They learned to go just so far from the water's edge and so forth.

After staying at the cabin Jack and I had built for a few days, I decided to go up the river about twenty miles farther. It would take me two days to go upstream, where I planned to stop for the winter. One afternoon a medium-sized black bear walked out on the sand bar just ahead of us. The dogs started after it and they nearly pulled the boat out of the river. My rifle was handy; when out alone I always have my gun in easy reach. I got one shot at the bear before it disappeared into the brush. I was sure I had hit the bear—this was the dogs' chance if they wanted to chase the bear so much. I unfastened them from the boat in about two minutes.

The dogs started to bark so I went to them to find out what was cooking. The bear was lying dead about one hundred yards from where it entered the brush. This meant bacon and lard for me. The bear was too heavy to drag to the beach, so I skinned him where he lay. I cut the meat in quarters and carried one quarter to the boat which was all it would hold. We were now happily on our way upstream again. I came upon an old abandoned trapping cabin. Magazines found in the cabin were dated 1901. With a few repairs made, such as chinking the walls with moss, a little fixing of the roof and door, one couldn't ask for a better place to put in the winter trapping. This was about sixty-five miles up the Salmon River. I had been there about a week when, glancing down the river, I saw a man leading his boat upstream. I didn't know who it could be until he poled his boat over to the cut bank side where I was located. Then I recognized him as a fellow I had seen in Fort Yukon.

I have been calling everybody "old." They *were* old there. Only about three of those oldtimers are left now. They have all died off—over forty of them are buried in the Old Hudson Bay Cemetery. They were a hardy bunch.

CHAPTER IV												Pulling Out

They lived tough, ate tough, and died hard. The real pioneers lived in a bygone age; we will never see their kind again. They thought nothing of poling or lining their loaded boats up some uncharted river, spending all summer on the way. These trappers were the men who opened up our last frontier. Next in line were the prospectors and miners, and then the missionaries moved in after the miners. The generations growing up now can't go ten miles without riding in an airplane.

Oldtimers at Fort Yukon: (Front row, left to right) Fred Schroeder, Pete Nelson, Bill O'Brien, Jack Warline; (rear) John Roberts, Lynn Shuman, Old Peterson and Bill Carney. (Courtesy Carroll family)

I had been settled for about a week when Pat Deagle arrived on his way up to his trapping cabin about twenty-five miles up from me. I invited Deagle to stop overnight with me, or for as long as he wanted to. He stayed two days and rested up. I told Deagle I would be up to visit him that fall.

The Salmon River was full of salmon, kings and dog salmon. The king salmon run was about over; the dog salmon run was just starting. The cabin I was living in had a fine spawning hole right in front of the cabin door. One could see the salmon by the thousands spawning in

65

the loose gravel of the river bottom. The water was so clear that it looked green. There were hundreds of grayling mingling among the salmon—stealing their spawn. The males would chase the graylings away and come back to take their place alongside the females.

I was in real bear country now and I used to see them even in the daytime close to the cabin. I commenced to think that, after all, the old boys down at Fort Yukon were telling the truth regarding the bears—grizzly and brown bears. By the signs one would think all the bear in Alaska were congregating on the Salmon River. There were two colors: gunnysack brown or dirty yellow. Sometimes the brown had silver tips along his back. I shot a few. I couldn't kill all I saw—there were too many of them and I might run short of ammunition. I had two rifles, a 30-40 carbine, an 8mm Mauser, and an old single-barrel 12-gauge shotgun. I hadn't killed a moose yet. There were no moose signs. The bears might have kept them back from the river. There wasn't much to do now until trapping started.

Malamute dogs are afraid of brown bears, but they will chase the common black ones. I have seen my dogs so scared of brown bears that they would run along the shore smelling toward the brush when, all of a sudden, they must have smelled a brown bear close, for they all jumped into the river and started swimming after the boat, which was a small affair manned by a pair of oars. I would pull the dogs into the boat. Of course, they had to shake themselves and get me all wet. A person shouldn't take his dogs on any scouting trip, especially loose dogs. Should they scare up a bear, the dogs will run straight for you with the bear after them

I've seen dogs, my own, too scared to bark. I went up the river one time to hunt. This time I took the dogs along and set up my tent and stove. I tied the dogs close to the tent to frighten away the bears that might come around during the night while I was asleep. About twelve

CHAPTER IV — Pulling Out

or one o'clock something wakened me; I didn't move, but listened, and I heard something down at the boat; still I didn't move; again I heard a noise at the boat—it sounded like an oar dropping in the boat. If it was a bear, why didn't the dogs bark? I knew there was fish blood in the boat from some graylings I had caught that day. Well, I thought, I had to do something. I eased out of my sleeping bag—my rifle was close to me all the time. I peeked through the tent flaps down at the boat. The night wasn't too dark with all the stars shining—I could see a big brown bear standing by the boat looking towards the tent. I took the best aim I could and let drive with my 30-40. I must have hit him. He reared over backward and let out a growl. I kept shooting in his direction until the rifle was empty. I had my old single-barrel 12-gauge handy should the bear change his mind and come back at me. In the meantime the dogs were howling their heads off. They were poor watchdogs. The bear could have come right into the tent without their making a sound. I took the tent down and put everything in the boat, including the dogs, as quickly as I could and got out of that section of Alaska. It was dark, but I could see the shoreline from the water. I floated down river a little over a mile to where there was a big driftwood pile I had noticed when coming up; I set the driftwood pile afire, and waited by the warm fire until daylight.

Some of the Alaskan freight dogs are afraid of bull moose in the fall. I was out across the river cutting out some trail where I knew there was some marten. I made, from canvas, two dog packs. The dogs packed what few things I needed. My bedding was a load for one dog; the other dog had in his pack a little tea and sugar, rice, a piece of bear meat, and dried salmon for the dogs to eat. I could always eat some of the fish myself for a change. A good-sized dog can pack 35 pounds all day. Some dogs can pack more than others, according to their builds. I went about six miles, brushing out some and blazing trees on both sides so I could see the blaze on the trees coming home. When out like this I always worked until dark (this made the night shorter), then I would make a campfire to cook by and for warmth. I had a moose bone with me; this is the shoulder blade taken from a moose, with the meat scraped off and the bone dried. This bone, when rubbed against willows and brush, sounds just like a bull moose hooking his horns at rutting time in the fall, calling another bull to fight with. If the night is calm they can hear this bone five miles away. Hours after you stop scraping the bone their sense of direction is so good they can make a beeline for the last place the noise came from.

    I proved this one night on that October trip. Before I crawled into my sleeping bag to go to sleep I scraped the moose bone for about ten minutes and then I listened for quite a while. This particular night I got an answering call from a long way off. I scraped the bone some more and listened for any sound I might hear. I could hear brush cracking this time, still a long way off, but getting closer all the time. At times I couldn't hear anything. The moose himself must have stopped to listen. I rubbed the bone again very lightly so the moose would think I was a long way off. He started breaking willows harder and getting closer fast. He was breaking down dry poles and sounded like a bulldozer coming through the thick woods. He would stop a few seconds to listen, rake his horns and grunt. Meanwhile, the two dogs lay still without a whimper. The moose was very near me now. I untied the dogs and told them in a whisper to go get him; instead, the two dogs curled up in my bed, scared to death. The moose was very close now but I couldn't see him through the thick brush. The reason I had turned the dogs loose was that I thought they would hold the moose until daylight. I have had oldtimers tell me how one dog had held a moose for them. I was afraid to use the bone any more for fear the moose might charge me. He was doing a lot of grunting. All at once he bolted through the brush and ran away. I could hear him for five miles. It sounded as if he were tearing down the whole woods with his horns, which, considering the noise he made, must have had a large spread. I gave the dogs a light switching for being cowards.

Sometimes a brown bear will fool a moose by clawing at a dry hollow tree. A moose, generally on the run, runs in the direction of such a sound; the bear charges the moose that has run right to him and hangs on until he gets the moose down. Sometimes a large bear grabs a moose by the neck and gets the moose down that way. The only evidence left is a huge set of horns. I once ran upon a place where a grizzly bear and a bull moose fought a battle. They tore up and broke down small trees over nearly half an acre. The bear always seems to come out the winner.

During the winter black bears crawl in their previously prepared dens and sleep for about seven months. The grizzlies are out later in the fall and wake up from their long hibernation earlier in the spring. It is very dangerous to fool around a grizzly bear den too soon after he holes up, or just after he wakes up in the spring. If you think you have a bear in a hole, never stand in front of it; he will charge out after you like a bullet. Black bears are not dangerous in this way.

On September 22, I started up the river to visit my friend Deagle. I intended to hunt moose along the way. I left the dogs home this time, throwing them some extra eats and leaving them water. There was an old washtub left in the cabin I was living in which had some holes in the bottom. I calked these up with some cotton rags to make it watertight. I filled the tub up with water and set it among the dogs. The tub full of water couldn't easily be tipped over; the dogs could reach it to drink even though chained. The dogs hollered their heads off when I was leaving them behind.

Above The Arctic Circle

I made it up to Deagle's in two days. My boat was practically empty. I camped out one night halfway to Deagle's. Not knowing the country well, I camped alongside a bear trail leading from a lake to the river. I guess the bears traveled over this portage every night, coming to the river to fish. The bear trail was eight inches deep in the ground from having been traveled over so much by the animals. I didn't notice this trail when I was putting up my small tent; it was pretty dark anyway—I saw the deep trail next morning. After I got the tent set up and a meal cooked and eaten, I crawled into my sleeping bag and tried to sleep. I blew out the candle about ten o'clock and finally dozed off to sleep. I was awakened by a sharp crack, like someone stepping on a dry stick just outside the tent. The first thing I did was to grab my rifle. I thought of shooting through the tent. The bear was making a deep blowing sound and a whining sound like a mule. It was pitch dark with a drizzle of rain. I stepped outside the tent with the rifle. The bear was still blowing and growling at me. He seemed to be standing up on his hind legs behind a large tree with his paws partly around the tree on the other side. I fired two shots at him. He never moved but kept on growling at me. With my old single-shot 12-gauge shotgun I gave him a dose of fine shot; this would move him, if anything would. The charge of shot must have hit him in the face. I could hear him running and breaking down brush for nearly an hour. I wasn't scared? I did not know what to do I was so frightened. I knew I could not sleep there the rest of the night, so I gathered up some wood and made a big fire which I kept going for the rest of the night. For courage I brewed some coffee in a can I had along.

The next afternoon I reached Deagle's. He was glad to see me. He told me that he had killed a big bear and a bull moose. I told him he was more lucky than I was. I hadn't killed a moose yet, and the bears were driving me nuts. Deagle, a carpenter by trade, had built himself a fine cabin. It is still standing today and is as good as the year he built it—almost fifty years ago.

I scouted around a couple of days with Deagle trying to locate a moose. We never saw any fresh signs. Whenever Deagle saw a fresh mink sign he used to get down on his hands and knees to smell it, and then remark, "That track must be fresh because it smells strong." There weren't so many bear signs up this way. Most of the good spawning holes were below there. When I was preparing to go back to my place Deagle gave me a whole quarter of moose meat. It must have weighed 150 pounds. On my way back I stopped to hunt several times back of the willow bars and on the islands where the moose usually feed. I wasted so much time hunting the second day that it was going to get dark before I made it home. I was just floating down with the current, making no noise, and not scaring any game away. Floating down a stream one has a chance to spot game easily. Going upstream is all work or you don't go forward.

It was getting quite dusky. I spotted a huge bear just ahead of me. He was fishing in the middle of the river, where the water was only six inches deep. I had my two guns ready. I was bearing right down on the bear. If I hadn't shoved the boat over with an oar, I would have butted right into him. I made a slight noise in getting the oar from inside the boat. The darn bear heard me and reared up; at about the same time I let him have a 12-gauge shotgun blast right in his belly. He reared and ran in circles, then towards the brush. He was probably almost blinded from the shot charge. I fired a couple of shots at him with the rifle, but missed; it was too dark to use the gun sights.

After I got home I did some fishing with a gaff hook and I hooked out five hundred dog salmon for the dogs to eat during the coming winter. Some of the fish I cut up to dry as I had seen the natives do down at Fort Yukon. A gaff is a piece of steel a quarter-inch in diameter, bent to the shape of a big fish hook and filed to a keen point. This hook is fastened to a pole about twelve inches long made for the purpose. With this hook, I could stand on the shore and gaff fish as fast as I wanted to. After the salmon spawn they stand guard over their buried eggs until they eventually roll over on their sides and the river current carries them downstream. They always drift close to shore. The river bars were lined with dead salmon. The stench of these decomposed fish must have drawn the bears for miles around. The bears won't eat a dead salmon as long as the river is open and they can catch live ones. They only eat the heads of the live salmon they catch; this must be the best part of the salmon.

One morning, just about daylight, the dogs were all whining as though they had seen game of some kind—most likely a moose. I got up and hurriedly stepped outside with my rifle. The dogs were all looking across the river at a bull moose standing on a bar. My first shot missed the moose. He just stood there looking across the river at us. The second shot got him back of the shoulder. That was the fatal shot. The moose stood a few seconds, then keeled over dead. This was the first moose I killed, and it was a big one. He could have been fatter, but the meat tasted better than bear or lynx.

I was well fixed now for meat during the coming winter. In a couple of weeks more the bull moose would be so rundown that they would not be fit to eat. The cows and calves don't seem to get in this rundown condition; they seem to stay in one place.

The first snow of the season fell on October 10. This was just what I wanted. I could go out now and scout around for fur signs. There seemed to be quite a few signs of lynx and mink tracks; no marten; they stay back from the river. I went back from the river about twenty miles with a packsack on my back, staying two nights in the brush. Marten signs looked good. This was in a westerly direction from my cabin, towards the Christian River.

CHAPTER IV — Pulling Out

I moved out with the dogs, taking my small tent and sheet-iron stove along; also fish for the dogs to eat and some grub for myself. Dried salmon makes good marten bait. I took out quite a load, including traps, snares, rifle, heavy axe and my sleeping bag on top of everything—all lashed down with rope. I traveled slow, brushing out thick places that I couldn't get through otherwise. I camped wherever night overtook me. I set traps as I went along. I started back on the seventh day away from the cabin. I had set traps for marten only; they are the first fur bearer to get prime in the fall, and they are the most curious and the easiest to catch in a trap also. They are the first to get unprime in the spring. In going back to the cabin I picked up eleven marten and three ermine, which got caught in marten sets. I was snowshoeing out a side line and setting traps as I went along. I carried the trap in my packsack and only went about half a mile, setting eleven H. O. Jump traps. The only reason I set the H. O. Jump Traps was that they were light to carry and to pack on one's back. The best trap for marten is the Number 1 or $1^1/_2$ Victor. In a matter of three hours, I turned back to the tent—my snowshoe trail was full of fresh marten tracks—three had been caught in my sets in that short time. It looked as if I had hit a real marten pay streak. "By all the signs," I thought to myself, "I may catch a hundred marten in this place!"

Some of the oldtimers down at Fort Yukon told me about the wolverine and how they follow one's trap line and rob it of anything that is caught; what they don't eat they hide so cunningly that it is hard to find. One time a wolverine stole a red fox from me that got caught accidentally in one of my marten traps. The wolverine didn't eat this fox. Instead, he dragged it out into the thick brush and cached it under the snow. Then he made a trail beyond the cached fox. He padded the snow down so evenly on top of the fox that one couldn't tell what he had done with it. It had me puzzled for a while. I looked up in all the treetops around, and I couldn't see anything. I figured the fox must be under the snow someplace along the trail. I took off my snowshoes so my feet would sink through the snowshoe trail and I would be able to feel any hard object under it. By doing this I found the fox undamaged.

The wolverine is a natural-born thief. He will steal anything. He is not a dangerous animal as some would have you believe. He eats anything he finds. He likes to follow the caribou herds and eat the leftovers from the wolves. During the night time they follow lynx and eat rabbits killed by the lynx and left uneaten. I've seen wolverine dig for mice in dry meadows. I've tried catching wolverine in different ways. You can leave a quarter of moose right in your trail and they won't touch it. It appears to them, "This is too easy—there must be a catch somewhere." They pass it by every time. But if you build a simple cache about seven feet high, more or less above the ground and put the same quarter of meat on top of this cache, cover it over with an old piece of canvas or brush; set all the traps you want to openly, he will climb right up this cache to steal. Of course, he will step right into the traps and get two or three of the traps on his toes. The wolverine can be classed as the toughest small animal we have, and the most active. He can run up and down a tree like a squirrel. He can come down head first or back up just as fast. Once caught, a wolverine never quits working to get loose. He chews up everything. His fangs are large as a wolf's. He is a mixture of the following: teeth like a bear, claws like a lynx, ears like an otter, with the stripes and scent of a skunk—and he only weighs around twenty to thirty pounds.

I stayed home about a week setting out traps for mink and snares for lynx. I went ten miles up river and ten miles down river. On making my second round to the marten patch I picked up seventeen marten and some weasels. I noticed a wolverine had visited my camp. I

| CHAPTER IV | Pulling Out |

couldn't notice anything gone that he might have stolen until I couldn't find my big double-bitted axe that I had left behind on my first trip out, as I didn't want to be hauling a heavy axe back and forth. I had another axe at my home cabin. I hunted for my axe every place possible—no axe. One is helpless out at camp without an axe. I started walking around my camp in circles—extending the circles each round. Finally, I came upon the wolverine's track coming from the direction of my camp. I followed his tracks; in one place it looked as if the wolverine were dragging a pole by the marks in the snow. I followed him about three hundred yards, and there lay my axe. Lucky for me he hadn't buried it. I would never have found the axe out there. I remembered I had cut up some bony meat with the dull side of the axe, which left some blood on the blade. The wolverine, thinking he had found something edible, simply packed my axe off. If it had snowed in the meantime I never would have found it. He probably heard me coming into the camp while I was a long distance away, and dropped the axe and ran away when he heard me coming.

During 1912 – 13 a certain "Sawbones" was appointed U. S. Commissioner for Fort Yukon. He was young and ambitious, and followed the dictates of his superiors. He subpoenaed practically the whole population of Fort Yukon, or about forty people in all, to appear in Fairbanks, Alaska, by December 5, 1912. Charges ranged from illegal cohabitation to land injunction cases. There were only about three cases involved; the rest were witnesses.

Chester Spencer, Deputy Marshal from Circle City, spent weeks serving subpoenas on trappers. Some were dragged in from their trap lines for miles around at a time when they should have been attending their lines. Spencer arrived at my cabin when I was out on the line. Luckily I came home that night or they wouldn't have known where to look for me with trap trails leading in all directions. Spencer had a guide with him from Circle

City named David John. I knew them both when I was in Circle during the winter of 1910 – 11. I got home early that evening. Before I left the cabin that morning I had left a partly frozen lynx wrapped up in my bed. This was to prevent it from freezing harder. I wanted to get the lynx thawed out so I could skin it for its pelt. Lynx have thousands of fleas on them even when the flesh is partly frozen. As soon as the cabin warms up the fleas leave the cat and get all over one's blankets and clothes. The Marshal got quite a jolt when I unwrapped the dead, frozen lynx from my bedding. David nearly laughed his head off. They both stayed with me that night. They told me in the morning that the fleas nearly ate them up alive and that they hadn't slept a wink all night from scratching. Well, I said they kept me awake all night too. After breakfast that morning I told them about an old trail leading across to Joe Ward's on the Porcupine River— about thirty miles across. I told them that I was sure I could guide them across, although we might have some trouble finding where the old trail left the lakes. Going over this portage we had to cross several lakes, but it would save us a hundred miles of travel. Otherwise we would have to go down the Salmon to its mouth, then up the Porcupine River forty-five miles. The Marshal had a subpoena for Joe Ward, Fred Schroeder, and me.

We navigated across country without any trouble. When we arrived at Joe Ward's, it was about dark. However, we all stopped with Fred. He had a larger place for us to stay. They were dumbfounded at seeing the Marshal; they couldn't think of any crime they had committed. They were slightly ill at ease when the Marshal handed them both their subpoenas. Their comment was they knew nothing, heard nothing, saw nothing. Schroeder cooked us up some lynx-burger steaks. We were so tired out, we didn't care what kind of steaks he cooked. They tasted fine. We kept asking Fred, "What kind of steaks are we eating?" He hesitated for awhile, but finally told

CHAPTER IV — Pulling Out

us they were made from ground lynx flesh. Whatever kind of meat they were made from, they still tasted good.

The next morning, Fred was up first and made a fire in both his small camp stoves. He mixed batter for hot cakes; he put a little bit of everything he had in his cakes, except beans. He cooked them fry-pan size over his two sheet-iron stoves. This would make them about ten inches in diameter, and possibly a half-inch thick. I know I could only eat half of one, while Fred ate two. It was funny to watch Fred prepare his hot cakes to eat. He would pick one from the container, hold it up to his face with both hands as if he were reading a paper. In this position, he would feel the hot cake all around as though he was feeling for a tack. David and the Marshal each ate one. There was a side of lynx, which neither ate.

We all left for Fort Yukon, seventy miles away, by portage winter trail. It took us two long days to reach Fort Yukon with light loads. Joe and Fred loaned me one dog for the trip. We rode part of the way where the trail was good. We looked like a caravan coming into town with five dog teams in a row. When we arrived in town everybody in Fort Yukon was preparing for the long trip to Fairbanks. There were roadhouses, and there was overnight sleeping for those who carried their own bedding and blankets. It took us three days to get up to Circle City, following the mail trail on the river ice. The mail trail from Circle to Fairbanks took us six days. We had a little difficulty in crossing some of the summits between Miller House and Faith Creek, which were always drifted in with snow. The creeks we had to cross were overflowed with six inches to a foot of water in some instances. We had to go through this water where there was no way of getting around it. The government allowed us fifteen dollars per day from the time we left Fort Yukon until we got back again. We had to pay all our expenses out of the fifteen dollars. Most of us stayed at the Pioneer Hotel, which was owned by Dave Petre. The hotel had rows of dog kennels to take care of the dogs. We attended court every day. Judge Fuller was the District Judge at that time. Harry Beancook's trial opened first. His was the most important case. Mrs. Pulrang, a witness for the prosecution, wore a band of white ermine furs around her hat; this made her look conspicuous in Court. She sat in front facing the Indian witnesses. When a question was put to a native through an interpreter by the defense, if the question needed a "yes" or "no" answer, the lady with the ermine hatband would, for "yes" bow her head up and down, and for "no" shake her head sideways. Beancook's lawyer, able Tom McGowan, caught on to Mrs. Pulrang. The judge asked the spectators in the courtroom if any of them had seen this lady nodding her head to the native witnesses? Several in court said they had seen the lady in the ermine-trimmed hat nodding her head at the native witnesses. Two steamboat captains stood up and verified the fact that the others had stated about this lady trying to intimidate the natives. This must have been very embarrassing for her. Old Tom McGowan, Harry Beancook's lawyer, dressed Mrs. Pulrang up one side and down the other. After this, the court recessed for a while, so Tom McGowan knew he had his client's case in the bag. There were about twelve witnesses still to be heard, so the trial dragged on a while longer. Before the case went to the jury their verdict was acquittal, of course. Moore's case was identical to Beancook's. As the jury had freed the "Big Shot" they couldn't do otherwise but free the "Little Shot."

Next came Clark's trial. He had been arrested for trespassing on Mission Ground. His case didn't last long; he was freed also. Clark was boiling with anger at having been dragged so many miles from the Chandlar River, over one hundred miles inside the Arctic Circle. He had to go home over unbroken trails, over mountain ranges, through blizzards, cold weather, sometimes fifty degrees below or lower, through uninhabited wilderness—all for nothing whatever.

We were in Fairbanks more than a month. Cost to the

CHAPTER IV — Pulling Out

government: over twenty-five thousand dollars with not one conviction. I, myself, was paid eight hundred dollars in twenty-dollar gold pieces. We couldn't all start for home the same day. Roadhouses could only accommodate a limited number of us at a time. We got quite a write-up in the local newspaper. I remember the heading was, "Fort Yukon Leaving Fairbanks, Bound For Home. We all Regret To See Them Leave So Soon."

The trail was worse going back, with overflowing creeks, drifted summits, cold winds, and snow. Many times we arrived at the roadhouses way after dark with our clothing wet from sweat, snow, and water. Our clothes were still partly wet the next morning. We had to put them on just the same and hit the trail. I had a passenger for Fort Yukon with me going back. This meant that I had to walk and run all the way. The only time I rode was when going down the hills and mountain slopes. To keep from going too fast downhill I used to wrap heavy dog chains around my sled runners, this held the sled back some. When the trail was heavy my passenger used to get off the sled and run awhile to keep his blood in circulation. Through the mountains the trail was about half uphill and half downhill. My passenger used to walk uphill. I let him hang on to the handle bars, this made it easier for him to navigate.

On our way back we stopped again at the Circle City Hotel, run by Mr. and Mrs. Fred Brentlinger. We still had three more hard days yet to reach Fort Yukon and home.

The payoff was that after all the money Beancook and the government spent, and all the hardships he went through on the trail, he married the girl that at one time was his housekeeper. He could have married her without going to court in Fairbanks for trying not to marry her. Although it may sound funny to a layman, we were all away for nearly two months—this didn't leave much time to do any more trapping that season. Nevertheless, I went back up to my trap line. I knew I had one lynx up there wrapped in my bed blankets to keep it from freezing too solid. But after two months gone away and some 50 degrees below zero or lower weather the lynx was solid as a brick. It took three days to thaw it out by the stove, after which I skinned it out for its fine fur, which might someday adorn some fair "milady"—but not if she knew how many fleas it had. The flesh looked good enough to eat; I didn't know whether to feed it to the dogs, or to grind it up into lynx-burgers as Schroeder had done. But I couldn't do that either, because I had no meat grinder.

It had snowed about eight inches during my absence from here. I couldn't find all my traps on account of the snow. No doubt I lost lots of fur by the animals having dragged the toggles off. I could only find about half the traps I had set before going on that nightmare trip to Fairbanks. That particular eager commissioner was retired of his commissionership and Fort Yukon has lived quietly and in peace ever since.

Time seems to go fast when one is interested in his work. The ice cleared out of the river about the twenty-fifth of May. In the next couple of days I would be floating down to Fort Yukon. For the season I had caught thirty three marten, twenty ermine, fifteen mink, fourteen lynx and the one wolverine I had outsmarted. Old Deagle came floating down from his cabin above me the day before I was leaving. The first thing I asked him was how his luck had been. He said that he did pretty good considering he had no dogs to pull his junk around. I made a lunch for the old man out of goose meat, coffee, bannock (one-inch-thick hot cakes cooked on top of the stove). I had three geese cooked up to eat while floating down the river.

Deagle was kind of slow in telling me what he had caught; but after he drank his coffee he loosened up and told me what he had caught, and it was good for an old man without dogs. He told me he had caught fifty-two

CHAPTER IV — Pulling Out

marten, nine mink, two wolverines, two lynx, and twenty ermine. He said he caught the marten on two short lines east and west of his cabin. He hadn't trapped much along the river except for some mink.

It took us six days to float down to Fort Yukon. The usual bunch of old trappers who had gotten into town ahead of us were lined along the bank of the river to greet us and shake hands all around. We unloaded our boats. Beancook had a log shack for me to stop in. The first thing I unloaded was my sack of furs. I took them directly to Beancook's store. The dogs came off next. Beancook was glad to see me and remarked, "It looks like you never slept much after getting back to your trap line, after losing the best part of the trapping season on that midwinter vacation!"

I paid Beancook what I owed him while in Fairbanks and showed him my furs. His offer was three hundred dollars. I told him, "The furs are yours, Harry, because you were so good to me last fall when you trusted me for an outfit to go out with when I was practically a stranger to you. I hope you make a profit on the furs. When you lose—it hurts all of us."

CHAPTER V
Trapping With a Family

I went up river to trap again, taking a partner along by the name of Roy Fox. I knew him from Circle City. We made it up to my trapping cabin without any incident. Fur prices had dropped below the previous year's. One afternoon we went for a stroll, thinking we might run into a bear or moose. We did run right into a big brown bear. He was nibbling at some highbush cranberries, which are different from those that grow close to the ground. They also have a large seed. This bear we ran into was one of that dirty-colored kind. He raised up on his hind legs. At that same instant, I fired. He was close, but he disappeared into the brush. I turned partly around to see what Roy was doing. He still had his gun to his shoulder, and was trying to pull the trigger with the safety on.

Roy said, "I'll bet you never hit the bear."

"He was so close, how could I miss him?"

We argued back and forth. Roy was convinced I never hit the bear; true, there were no blood signs.

In our arguing we both nearly stepped on the bear—stone dead. He was hard to see; his coat of fur blended with the dry grass and bushes. He was lying in sort of a hollow.

I told Roy, "You had better fire a shot into his head to make sure he is dead."

So Roy shot at his head and missed by ten feet.

I said, "Shoot him again, Roy." My shot had hit the

bear, who was turned straight toward me, high up on the breast, going through the heart, and out on one side of his backbone. At that, he had run a hundred yards. We never caught much fur that season. Roy's worst trait was wanting to go down to Fort Yukon every time we had a couple of furs on hand.

He used to tell me we should sell our furs as fast as we caught them just in case the market dropped. Roy was about my age. His home was at Circle and Fairbanks. I knew the real reason why Roy wanted to go into town so often—it was to see his girl friend. I never disagreed with him about making these frequent trips to town, because I had a girl friend myself down at Fort Yukon, who later became my wife. Selling most of our furs as we caught them didn't let us accumulate many furs to take down with us. We floated downstream the usual way. We didn't stay long in town—about a week. Roy wanted to get to Circle City as soon as he could and I was going up to put the summer in out at the mine. We started up the Yukon River. I wanted to make a stake to get married on in the fall. We followed the mail from Fort Yukon to Circle, leaving April 20 with the dog team. We rode most of the way up. We made the eighty miles in three easy days, as the trail was in fine shape. I left my dog team in Circle and rode out to the mines with my good friend, Nels Rasmussen. I got a job cooking as soon as I got out to Miller House. The name of the man whom I worked for was Frank Miller—a fine fellow; he still lives at Miller House. I haven't seen him for forty-one years. We cooked and ate in a big tent on Porcupine Creek.

The caribou were so thick out there that the hills and mountains looked like a moving mass for days; there must have been tens of thousands. Today I don't think there is 10 percent of the herds left. One herd, on their annual migration, used to cross the Yukon River above Circle City. This was a large herd also. At times, there

CHAPTER V   Trapping With a Family

were so many caribou in the river that the steamboats had difficulty navigating through them. They would nearly choke up the River Channel. For fear of damaging their stern wheels, the captains sometimes tied their boats up to let the herds pass on their way, or to thin out. No caribou have crossed at this place for many years. The reason? There are none left to cross. It would be hardly fair to blame the game wardens for this disappearance. There was no game warden at that time. I believe lots of the caribou died from disease, or had been slaughtered by the Eskimos for their hides and tongues in the barren lands of northern Canada.

I cooked for Mr. Miller all summer. I quit about September 1. Mr. Miller didn't like the idea of my quitting before the end of the season; as usual, I could have worked another month. I explained to Miller that I was getting married down at Fort Yukon to a wonderful girl, and that I had a long boat trip to make before the freeze-up. When I got back to Circle City I gathered my dogs and bought another one; then I bought one of those square-stern boats. Also, one five-gallon keg of whiskey. I paid twenty-five dollars for this whiskey—this was to celebrate the wedding.

I arrived down at Fort Yukon September 14, 1915, and was married on September 15. The old Indian, Rev. William Loalo, married us. The church was crowded. I felt embarrassed. The old native minister could only speak in his native language. Lizzie Wood, the missionary previously mentioned, stood up for me and another old friend, Frank Foster, was best man. For all I knew I might have been marrying Lizzie Wood—seventy-two years old—instead of Fannie Martin. I couldn't understand a word of what the Reverend was saying.

After the wedding was over I walked out of the church a married man. It was then that I dug up the five gallons of whiskey. Later, all the whites drank their fill, including the marshal, who liked to be called "Dock." Dock sure liked his drinks. He used to wrap his thumb and forefinger around the top of his glass; these acted as sideboards. He could double the volume of his drink this way. Dock claimed he was the son of a minister, but nobody believed him. After sizing up the crowd I decided Dock was the soberest one of the bunch, so I put "him" in charge of what was left of the five-gallon keg. This may sound funny after all he drank. Dock was to see to it that no native got hold of any of it. It was unlawful for the natives to drink at that time. Dock didn't spend much time watching the keg. He simply picked up the near-empty keg and went home with it. It was a few days before Dock showed up again. I stayed with the gang until the wee hours of morning. The party was being held at Jim Haly's famous roadhouse. Fred Schroeder, the songbird of the party, did most of the singing. Once Fred got started singing nobody could stop him. Someone in the crowd threw an egg at him and it hit the poor old fellow square

James and Fannie were married at St. Stephen's Church in Fort Yukon on September 15, 1915. The Indian Reverend William Loalo conducted the ceremony in his Native language. (Dr. Cook Collection, UAF 2003-109-110, Archives, University of Alaska Fairbanks.)

CHAPTER V Trapping With a Family

in the mouth. Of course, Fred wanted to lick the fellow that threw the egg. Nobody knew who in the crowd had done it. With some sympathy from the crowd, they got Fred singing again with promises that there would be no more egg-throwing; so Fred received no more eggs that evening. When the crowd wasn't dancing they were singing also—joining in at times with Fred. Apparently a good time was had by all.

I bought a good poling boat, but we still had to use track-line pole and dogs for locomotive power. After spending a few days gathering in supplies and equipment to trap with we were ready to go. This was nothing new to Fannie, as she was born in Fort Yukon. We took an old native up with us to bring the boat back to Fort Yukon, as we planned to come out by dog team the next spring. There was nothing exciting on the trip up except hard work and lots of bear tracks in the snow. Sometimes the ground is covered with snow two or three weeks before the rivers freeze over.

The native who helped us was called Peter Moses. Many years ago he moved up the Porcupine River, where there is an Indian village at the mouth of the Old Crow River, Yukon Territory, Canada. Moses was elected Chief of the Old Crow Tribe of Indians, commonly called the Rat Indians, so named for the tens of thousands of muskrats they take from the Crow flats and lakes each spring. Chief Moses was decorated by the king of England during the first World War.

The Chief never stepped ashore without his rifle. We saw some moose tracks, but never caught up with any. Nearly every night we camped out we could hear bears cracking brush on their way to the river to fish after sleeping back in the brush all day. This used to worry the Chief. He always slept with his rifle handy. We reached our cabin on the seventh day after leaving town. We hurriedly got the boat unloaded so Moses could start back as soon as possible before the ice started to run, which could be any day from then on. We still had a small boat with us to hunt and run around with.

Fannie and I went up as far as Deagle's cabin. I woke up just at break of day and slipped outside to listen. I could hear something heavy walking in the loose gravel of the river bar. It could only be a moose. I stepped back in the cabin and got hold of my rifle and I went out again. I could see two moose standing on the opposite gravel bar about two hundred yards away; just the right distance for a good shot. One of the moose was a large grayish one. The smaller bull showed up dark alongside the lighter one. I took a shot at the large bull which was standing broadside. He never made a move. I shot a second time; and a third time as the moose disappeared in the thick brush. During all the commotion the dark-colored moose never moved. We only wanted the Big Bull, if we could get him. So we started over to the spot where the moose disappeared in the brush. I had hit him—there was a trail of blood. By the way he was bleeding I knew the moose was ours. He was lying down when we came upon him. On sight of us he stood up but made no attempt to run. I shot him through the neck—afterwards we cut the moose's throat so he would bleed well.

We guessed the moose weighed about seven hundred pounds dressed; this is considered a large moose for this section of Alaska. The moose down at Kenai get much larger. We had a whole day's work ahead of us skinning and butchering the moose. We made a lunch before we started working on it. The dark moose lingered on the same gravel bar all day. Next morning he left for parts unknown.

The weather was so nice, we wished we could stay at least a whole week. The moose was hit two times, once through the horns—a wild shot, one miss, and one back of the shoulders—the killing shot. We packed the meat to the boat next day after putting some small willows in the boat. First we loaded in the meat; it made a good load for a small boat.

CHAPTER V   Trapping With a Family

First thing we did after getting home was to water the dogs and feed them. We had brought six dogs up with us. In the next few days we gaffed or hooked out of the river about eight hundred dog salmon. This was enough to feed our dogs all winter. We hooked these fish out from the spawning ground right in front of the cabin. We planned on running a hundred mile trap line that winter, going in a circle; this would eliminate backtracking. We had brought a thermometer with us so we could tell how cold it was. If it got below thirty-five below zero we wouldn't go out on any trips. Heretofore, I used to tell how cold it was by my breath; at forty-five below zero one could tell by one's breath cracking; at sixty below and colder one could blow a blue shaft of air that sounded like blowing against a piece of paper.

On our third trip around the trap line we were marooned for five days on account of the cold weather. We were camped in a small seven-by-seven silk

Taken in 1916, the woman at left is Fannie with her mother, May Martin, right. Fannie is holding her little brother Richard Martin born in 1915. Tallest child is Fannie's sister Abbie, partially hidden child is Fannie's sister Blanche. A special occasion with everyone dressed up. (Courtesy Carroll family)

tent; the weather must have been 60 degrees below or more. We had not brought the thermometer with us for fear of breaking it on the rough trail. All we had for a

89

stove was two five-gallon kerosene cans telescoped together. One end of one of the cans acted as a door. On the top of the back can we had a small stovepipe leading out. This pipe also was made from tin cans. When traveling, this stovepipe fitted inside the stove, thus saving space in the toboggans. This makeshift stove was a real freeze-out. The wood had to be cut fine and it kept one constantly busy feeding the stove with this fine wood. A big stick would about kill the fire; then it would get cold in a few minutes. During the daylight hours I did nothing but cut wood for the night.

This place was more than one hundred miles inside the Arctic Circle. Our grub and dog feed was getting low. The weather was far too cold to travel in. But we couldn't stay here until our last bit of grub was gone. We were now down to rabbits and rice. This would be practically all gone by the time we got home, which was sixty miles away. We made a fire on the ground from what was left of our split wood. This fire was for warming our hands while we were taking the tent down and lashing our toboggans. We took it easy for the first few miles so as not to get up a sweat when first starting out; otherwise, it would be hard to keep warm towards evening with our clothing damp from sweating. The only reason we didn't freeze to death was because we were young and tough, I guess. We each had a caribou coat to wear; there is nothing warmer.

The day after we got home the weather moderated to thirty-five degrees below, which felt warm after spending over a week in the extreme cold snap. All fur-bearing animals hole up except lynx, wolverine and timber wolf which seem to be at large in any temperature. We made a good catch of furs that winter, better than fifteen hundred dollars worth, big money in those days. The price of furs had advanced quite a bit in the past two seasons. We could buy provisions at that time for less than a third the price of what they cost today.

CHAPTER V	Trapping With a Family

Our catch for the season consisted of forty-six lynx, twenty-three mink, seventy marten. The lynx and mink were caught along the Salmon River, while the marten were taken back from the river in the flats and rolling ridges and drains.

It was now the twelfth of April, and getting to be time to pack up our fur and get started down the river to Fort Yukon. Luckily, we had a large-size toboggan. It was sixteen inches wide and nine feet long, and made of hickory boards twelve feet long and four inches wide. The stores those days sold this hickory and also maple

From left to right, Fannie's mother May Martin, Fannie's sister Abbie, Margaret Thomas. The ladies are drying salmon during a fishing trip to Old Rampart or Rampart House in this undated photo. (Courtesy Carroll family)

boards to the trappers to make their toboggans with. I used to make my own. I had a form to bend the boards on after they had been steamed. For a steamer I used a five-gallon oil can and two stovepipes. I fitted the pipe in one end of the can, filled the can full of water, and set it on a campfire to boil. I shoved the four boards down the stovepipe to the bottom of the can; I stuffed some gunny sack in the cracks around the top of the stovepipe to keep the steam in. In about three hours one could bend the steamed ends like rubber. After the boards dried on the form I fastened them with cross pieces. I made a basket from a partly tanned big moose hide—the full length of the toboggan. I had lash loops all around the top of the basket to lash the load down. We loaded the lynx in the toboggan first (they were the most bulky); the marten and mink one could get in a big gunny sack. We had a top-heavy load all right.

Where the trail was bad I used to snowshoe ahead, which made it easier for the dogs. Fannie hung on the handle bars or sat on top of the load which helped to press it down. The weather was fine. We experienced no difficulty in getting to Fort Yukon. I stepped through the shell ice once and got wet to my knees. We always carried an extra change of wool socks in case this happened. The first day or so after arriving in town we stayed with Fannie's mother until we could arrange for a place of our own.

During 1916 – 17 the whole country was alive with lynx and rabbits. We caught 225 lynx ourselves. The record catch was made by Joe Ward. Other big catches ran around two hundred lynx. If we had more traps and snares we could have caught another hundred lynx. In one day, trapping about fifteen miles, we caught forty lynx. We had to stack them up along the trail that year. We ate lots of lynx. Moose were scarce. Lynx meat tastes good if it weren't for the thought of eating a cat. The wolverines got away with lots of lynx on me until I got on to their habits of caching them in the snow.

There was a ridge at one place along our trail where we had trail sets made. The lynx always hit this ridge and followed our trail. I remember one time we were taking a lynx out of a trap on this ridge, and we could look ahead and see three more lynx already caught. There were just as many lynx tracks when we picked up our

CHAPTER V   Trapping With a Family

lines in the spring as there were when we had set our lines out in the fall. Nobody seemed to know where they all came from. They seemed to be migrating through the country from northeastern Canada. The next year I caught ninety-nine lynx. I tried hard to get one more lynx to make the even hundred, but I just couldn't. The run of lynx came in from the west and hit the west banks of the Salmon. It seemed I caught nearly all of them before they crossed to the east banks.

We had a whole boatload of lynx coming into town in 1917. Two hundred and twenty-five lynx is some bulk. We only skinned and stretched a few of the lynx as we caught them. After the trapping season was over we thawed, skinned, and stretched lynx for a whole month towards spring. As spring broke the lynx started eating one another. I caught a lynx close to the cabin. Something had eaten it in half. I reset the trap and the next morning I had a lynx caught. This proved they were running short of food and turning into cannibals. After this we never ate any more lynx.

We sold our entire catch of furs to Harry Beancook, the trader. He paid us $8 each for the lynx, $6 for marten, and $4 for the mink, or a total of $1,828. With what we saved from this big catch we bought a log cabin, an outboard motor that wouldn't run, and still had five hundred dollars left to pay for another outfit.

Before we left town the next fall we had Beancook order an outboard motor for us from the States. It was called Bronko, and made by the Bronko Motor Company. It was a one-cylinder affair with about a six-inch piston stroke. I don't remember the horsepower, probably a four or six hp. It must have been one of the first motors that company made and put on the market. It had a plunger pump that would wear out in less than a day's running in the muddy Yukon River. It was built rigid; by this I mean, if I should hit something solid, such as ground or rocks, the motor would either break in two

93

or pull the back end of the boat off. I fixed this trouble by strap-hinging two pieces of two-by-fours together. I fitted this to the back end of the boat with the motor resting on the top of the two-by-fours. I could now raise the motor up and down as desired. I should have taken out a patent on this and sold it to the Bronko Motor Works. I probably missed the chance to make a million. When the motor did run it used to shake the boat from bow to stern. Everything in the boat shook and rattled. We used the motor one season. The second season we tried to use the motor it refused to work. I couldn't start

James A. Carroll in 1928, freighting up the Porcupine River. Boats at the rear are power boats. Note the dogs on board in foreground. (Courtesy Carroll family)

it. We carried along a special piece of canvas so as not to lose any of the parts. Then we would take the motor all apart and look inside of it. This didn't do us any good, because we didn't know one part from the other. So I told Fannie we had better throw it ashore and go on with the track line, pole, and dogs. This we did, and I was glad of it. My hands were black and blistered from turning the flywheel, trying to get the thing going. The next spring we sold the motor to a fellow named Gordon, a trapper on the Coleen River. He still traps there

CHAPTER V Trapping With a Family

today. He told me the next year that he had no end of trouble trying to keep the thing going. He said he took it apart on every sand bar on the Porcupine and Coleen Rivers. "Finally," he said, "I was so disgusted that I threw the motor overboard. I even wore out the piece of canvas that I used to take the motor apart on."

During the fall of 1917 I trapped alone at the same place on the Salmon River. I left Fannie in town with her mother. I made a trip into town during the last part of December. It was then I learned that I was the father of a son born November 20, 1917, named Clifton. I now had more responsibility, which meant we would have to extend our trap line a few more miles. I had brought in a fair bunch of assorted furs, mostly marten. By this time we had a few dollars ahead.

We left Fort Yukon the first part of January, 1918. There were three of us going back now. To keep the baby warm Fannie made a rabbit-skin bag to protect him from the cold. Fannie was now tied down to the cabin, the baby being too young to take on the trail. So I ran the trap line myself as I had for two months. Sometimes I would be gone for two weeks at a time. This was a long time for Fannie to stay alone, but she didn't seem to mind it. I probably worried more than she did while I was away. After the ice cleared out of the Salmon we floated down to Fort Yukon. We had a fairly nice cabin in town to live in.

Bishop Rowe, Bishop of Alaska, whom everybody from Point Barrow to Ketchikan knew, loved, and respected, wanted someone to row him down to Tanana, or Fort Gibbon, as it was called in those days when Uncle Sam had soldiers stationed there. The Bishop singled me out to take him down in my boat. Rowing, the trip down would take five or more days. The Bishop was to send my boat back upstream on one of the many steamboats plying up and down the Yukon River before the railroad was built; and also pay my fare back to Fort Yukon and fifty dollars, which sounded good to me. The trip was a pleasant one. Nice sunshiny days. There were no storms or heavy winds, which would have made such a trip a miserable one in an open boat. The Bishop made brief stops at all the native fishing camps along the way. When meal hours arrived the Bishop had me pull ashore. He always did the cooking on an open campfire. I would gather up the wood for him. He preferred cottonwood, if obtainable, as it threw fewer sparks than spruce wood.

I didn't tell the Bishop that cooking out of doors or indoors was my profession. The Bishop did such a good job of cooking himself that I never interfered with him. After a few days spent at Fort Gibbon the Bishop decided to go back to Fort Yukon on the steamboat with me. My rowboat was put aboard one of the barges the steamer was pushing upstream. When we arrived at Fort Yukon all the Bishop's friends were on hand to welcome him back, and ask him how he enjoyed the trip. Someone said to the Bishop, "You had a good cook along, so you must have eaten well." I felt kind of sheepish for having let the Bishop do all the cooking, but he had enjoyed doing it. The Bishop turned to me smiling and said, "Jimmy, I didn't think you would do that to me."

Fannie and I decided to lay off trapping for a while and go in the wood business. A woodman in Fort Yukon by the name of Jerry wanted to sell out his wood business. He had two horses, sleds, and a power saw, etc. He had a two-story log house also, and quite a bit of wood on hand. He was asking $2,000 for everything so I and a green Irishman not many months from Ireland, Johnny Michal, formed a partnership and bought Jerry's wood business out, each of us putting up a thousand dollars.

Johnny didn't know much about horses. He couldn't even harness one. I was born on a Minnesota farm and

CHAPTER V — Trapping With a Family

knew a little bit about horses, as we had kept horses on our farm at home. Johnny wouldn't let me touch the horses. He figured they were the only two horses in Alaska, and I was too young to know anything about horses anyway. One day Johnny drove the horses hitched to a makeshift wagon down to meet the boat. He actually had the harness on upside down. He didn't know the difference until someone attracted his attention to it.

I went into the barn one day to see what Johnny was doing there so long. He had the harness on one of the horses turned inside out. He told me he had spent over an hour trying to untangle Barney's harness. Barney was the horse's name. I took the harness off and turned it for him. There was about three inches of leather hanging out on both sides of the breeching which was used to take up, or let out, the breeching. Johnny asked me what those two short ends were there for. I told him, "That don't mean nothing."

"Oh, but it does, or they wouldn't be there," he insisted.

One morning, Johnny came over to our house with a worried look. He was worrying about getting the horses shod. He said it would cost us fifty dollars.

I said, "The two horses ain't worth fifty dollars."

This used to get Johnny's goat.

About a week later, Johnny came over to our house again, worried-looking as usual. "These frosty mornings, the horses must have shoes on or they will be freezing their feet," he said.

"How in hell is a piece of iron going to keep their feet warm?" I asked.

"Oh, Byjases, horseshoes are made from the finest steel."

"You can get the horses shod if you want to. I am not putting any more of my money in those horses."

Johnny answered, "I will buy your half interest out, give or take."

So Johnny took and I was out of the wood business for good. So Fannie and I hurriedly got an outfit together and went up the Porcupine River as far as the mouth of the Salmon River. An old guy by the name of Dad Moore, recently down from Dawson, took us up to the mouth of the Salmon on an old skiff he brought down from Dawson. This was about September 28 which was late for the Porcupine River to still be free of running ice. The day after Dad left us at the Salmon mouth, the Porcupine River was running full of ice.

After the Salmon froze up solid enough to permit travel on it, I made a dog-team trip up to our trapping cabin on the Salmon River to bring down a load of traps and snares.

Dressed in their winter fur garments, Fannie poses with her mother, May Martin, in front of a boat. (Courtesy Carroll family)

We caught a nice bunch of mink at the mouth of the Salmon and we fixed up a little old cabin about ten-feet square, which was originally built by Chief Moses. The two babies (our family had grown) had to sleep on the table at night. This table was made from small poles hewed smooth on one side. After we got up in the morning we transferred the babies to our bed to give us the use of the table. The babies never got cold. Their mother made them each a rabbit skin bag to sleep in. We stayed until the lakes and rivers all opened up. By that time we had caught

CHAPTER V Trapping With a Family

and shot sixteen hundred rats. I did the shooting and trapping while Fannie did the skinning and stretching.

I had to make a cross-country trip up to our camp to bring our boat down to the mouth of the Salmon. We had to have this to float down to Fort Yukon.

Although the walking was tough going over swamps, "nigger-heads," brush, and water, I made the trip in two days. The only game I saw was a cow moose with her newborn calf. I thought she was going to charge after me. But she only stood still with her hair standing straight up. If she had charged me, I would have had to shoot, which I didn't want to do. I also ran across a duck's nest plumb full of eggs. I had a can I used to make tea in, so I put all the eggs in the can and put the can in my packsack. I packed the eggs about seven miles. My pack seemed to be getting heavier so I decided I could carry the eggs easier inside me than on my back. I boiled them all in the tea can, but they were a little too ripe for me to eat. It never entered my head to test them before I packed them so far. I stopped one day at our camp to caulk the open seams so the boat wouldn't sink going down the river. By rowing hard I made it to the mouth of the Salmon where the family was camped in two long days. From there, we floated down the Porcupine River to Fort Yukon, where we spent the summer.

Summer photo of James A. Carroll and three of his children, most likely Joe, Annie, and baby Mabel (in white), circa 1925 - 1926. (Courtesy Carroll family)

99

We bought a second motor from a fellow who came down from Nation, which used to be a small settlement years ago below Eagle, Alaska. It was a one-lunger with, possibly, a six-inch stroke. I had to strengthen the bottom of the boat to prevent the motor from pounding its way through. It was a four horsepower heavy-duty with a maze of wires and small batteries to keep it running. The fellow I bought the motor from had the base of the motor all puttied up with a yellow soap. I asked him what the soap was for. He said it kept the motor from sucking in too much air. I didn't know what he meant by that. I paid the guy two hundred and fifty dollars which was the price he was asking. I thought the price was too much. Anyway, we were getting tired of lining our boat upstream with dogs, rope, and pole. The fellow we bought the motor from guaranteed it to run after the engine was installed.

Summer in Fort Yukon, from right to left, Annie, Mabel, James, and Tommy (born 1927). Note the children are wearing beaded moccasins. (Courtesy Carroll family)

Our boat was thirty feet long with a shovel nose, the type most trappers used for going upstream to their trapping camps. I scraped off all the soap from around the

CHAPTER V   Trapping With a Family

base of the engine and cleaned up the motor in general. This particular type motor was supposed to take in a certain amount of air through the base which the owner had tried to stop up with soap. I connected all the loose wires to the engine and batteries for a trial run. The motor worked perfectly; in fact, I had a hard time to stop it after it started. I pressed every gadget that was supposed to stop the motor before I succeeded. I had a universal joint outside the stern of the boat which allowed me to raise the skag in shallow water.

We were going to put in one more winter trapping up the Salmon River—traveling in style this time. On our way up the Porcupine River to the mouth of the Salmon

James A. Carroll, "Up on the trapline on the Black River." Notice Carroll's dogsled has long handlebars. This made it easier to walk behind the sled while wearing snowshoes. (Courtesy Carroll family)

we could see an old trapper sitting on a log having his lunch. As we knew him well, we pulled in to shore.

He was on his way up the Black River. He had trapped for many years about a hundred miles from its mouth. He was using track line and pole. He trapped without dogs, making small fur catches each season. The old man warmed up a cup of tea on his campfire.

After we drank the tea I said, "Pete, you had better tie alongside us and we will tow you up to the mouth of the Black River. I have plenty of power to do it."

101

"Oh, no, no!" said Pete. He declined our offer; he said we would shake his boat to pieces.

Of course, I laughed and said, "I don't think so, Pete."

"Why," he said, "I could feel the ground under me shake as you were coming up. I even heard you when you left Fort Yukon." This was about eight miles from town. The one-lunger did make things shake all right, but it seemed to have power even if it did shake everything in the boat. We could run all day on five gallons of gas, making twenty-five miles a day. We made it home on twenty gallons of gas. After we arrived at our camp, we followed the usual routine.

In the late fall we killed many bears and a few moose, hooked out salmon from the river, caught so many furs, fell through the ice so many times without getting drowned, and in June, 1919, we had another addition to the family. We called him "Joe."

Dressed for winter in Fort Yukon, the Carroll children from oldest to youngest Joe, Mabel, and Tommy. (Courtesy Carroll family)

During 1919 – 1920, muskrats were bringing four dollars each on the fur markets. We decided to go to the old flats to catch 'rats. The Old Crow flats are located in the Yukon Territory, Canada. They are about seventy miles across and contain several hundred lakes, nearly all of them 'rat lakes. Some of the lakes are ten

CHAPTER V — Trapping With a Family

miles in extent. We decided to go to the Crow flats and make our fortune catching 'rats. The Crow flats are located about two hundred and fifty miles north of Fort Yukon by dog team. We would have to double-trip all the way with two dog teams in order to take along plenty of grub, traps, and stretchers. This would mean a trip of five hundred miles with double-tripping from Fort Yukon.

We had two toboggans; one ten feet, and one seven feet, long. Fannie would use the smaller one with four dogs and haul the babies, bedding, stove, and tent. I was to haul dog feed, grub, traps, etc. My toboggan was sixteen inches wide; Fannie's was fourteen inches wide. I also hauled a large canvas to make a scow to float down the Crow River after the ice went out. When we were planning this trip a lot of the oldtimers said we would never make it, and if we did make it we could never stand the climate. It was a tough country all right—almost treeless, snow blowing and making drifts so one could hardly cut into them with an axe, and blizzards that blew snow through tents. But we figured if trappers had made it in years past we should be able to stand it too.

Taken in 1920 on the trip to Rampart House, with Clifton and Joe bundled in their rabbit-skin bags, sleeping peacefully in Fannie's toboggan. (Courtesy Carroll family)

CHAPTER VI
The Crow Flats

We finally got started on our long trip to the Crow Flats on April 1. I had already moved six hundred pounds of our stuff as far as Shuman House, which used to be a small Indian settlement seventy miles from Fort Yukon. I used the seven dogs and my large toboggan with this load. I went along quickly, making the trip in four days. I had let the dogs rest one day before leaving Fort Yukon. Our loads were not too heavy.

From Shuman House we moved to a place called Old Village, another deserted Indian settlement twenty miles from Shuman House. Each of these way-places had at least an old tin stove in them, left there by trappers who traveled up and down the Porcupine River. Their windows had been broken years before. Stopping in one of these cabins was just a bit better than camping out in the open. Fannie and the two youngsters stayed in one, while I went back with all the dogs to Shuman House, bringing forward the goods we had left there. We tried after that to make it a practice of not making over fifteen miles a day. This would allow me to make a round trip the next day. We had to set up tent and stove when we stopped if no old cabin was available. The next deserted village was called Burnt Paw. This was over twenty miles above Old Village. This made a long day with heavy loads. Our next stop was Curtis's Place, commonly called "The Howling Dog" because the wind is always blowing through a narrow canyon just above Curtis's camp. It

105

makes a noise just like a dog howling. The distance between Burnt Paw and Curtis's Place was thirty-five miles. It took us three days to make it. Curtis trapped at Howling Dog for many years. We stayed with Curtis overnight. He treated us fine and, as he was a great talker, we didn't get to sleep until midnight.

Our next stop was to be Old Rampart, only fourteen miles upstream. We made this distance in a few hours. The weather so far was fine, although the nights were cold—thirty or forty degrees below zero. We stopped a couple of days at Old Rampart to give the dogs a rest. Old Rampart used to be quite an Indian Settlement years ago. The Hudson Bay Company was at Old Rampart once, thinking they were in Canadian Territory. Eventually they moved back up the Porcupine River some thirty-five miles to Rampart House where the Canadian-American lines parallel one another.

We stopped over a couple of days, as I said, to give the dogs a rest. We bought two bales of dry fish from the natives, which was about a hundred pounds at forty cents a pound. Our next stop was at Campbell River, fourteen miles from the Canadian line, where two partners lived and trapped together for over thirty years. We stopped with the partners for three days. In all those years they had never quarreled once. Hank was always willing to do all the work, and Henry never interfered with him. While staying with the partners we moved all our goods up to the Canadian line. It so happened that the partners were going to the Crow 'rat lakes too. We all decided to go out together. They had been out to the Crow Flats themselves the past couple of springs. They both said they knew the pass across the mountains, which meant everything.

The partners were always short of wood and never kept enough on hand to start the morning fires. I've seen Hank wake up at 5 a.m., pull on his parka and go up a timbered ridge back of their cabin trying to find a dry

CHAPTER VI   The Crow Flats

pole to start a fire in their heater, with snow so deep he had to use snowshoes to navigate, and with the thermometer hovering around forty below. If the fire went out in the heater for want of wood, it would get just as cold in the cabin as it was outside. There were holes in the roof of their cabin so big that you could see the stars shining through. Eventually Hank would get the fire going; next, he would fill up a couple of tin buckets full of snow and set them on the flat-top heater to melt for water to brew coffee. They were both great coffee-drinkers and cigarette-smokers. I got up from our bed on the floor and dressed. I didn't like to see poor old Hank doing everything. I thought I could help in some way with the cooking. They wouldn't let us use any of our groceries. Everything we ate had to be theirs—they were that kind. As mentioned before, this place where the partners trapped from was known as Campbell River. They did a little trading in groceries which they brought up by boat in the fall.

The two babies were standing the trip well so far. Our next stop was Rampart House, just inside the Canadian border, fourteen miles from Campbell River. We had to go through the customs there. There were two Royal Canadian Police stationed here. We paid custom fees on our grub, ammunition, and tobacco. The ammunition and tobacco carried a heavy custom fee. We also paid a five-dollar fee on each of our dogs. This fee was to be refunded to us when we brought all the dogs back with us. The police had a station at the village of Old Crow, which was situated at the mouth of the Old Crow River. This station was to contact the 'ratters floating down the Crow River with their catches, which sometimes totaled 60,000 or more 'rats.

At Rampart House we stopped with Daniel Cadzow; he had a fine house and general store. Cadzow invited us to stop with him in his house. We had known Cadzow for several years. He certainly made us feel at home. He fed us with the best he had. We bought some flour, sugar, coffee, and dried fish from him, making sure we would not run short of anything. This was our last chance to stock up before starting over the mountains to the Flats. Cadzow operated his store in the Old Hudson Bay Company fashion; he sold everything by the cupful. I forget

Daniel Cadzow's house; at the time it was the most expensive home in the arctic. Photo taken on the 1920 trip to Rampart House. (Courtesy Carroll family)

James and Fannie posed for a photograph at Rampart House in 1920 while en route to the Old Crow Flats. Notice the toboggan in the background. (Courtesy Carroll family)

how big the cups were. Sugar was 50 cents per cup; flour, 50 cents per cup; rice, 50 cents per cup; raisins, a dollar per cup; bulk tea, a dollar per cup; bulk coffee, a dollar per cup. Cadzow was a fine host, but he never had a heater in the store. When anybody wanted to buy something he had to dress up in his fur parka to keep from freezing.

CHAPTER VI The Crow Flats

It was a little early yet to start across the mountains. The partners stayed at the police barracks where one of the dogs bit our small son, Clifton, on the hand very badly; he carries the scar to this day. We didn't know what to do for a while. The police carried some first-aid medicine, such as iodine, salve, and bandages. This wound delayed our leaving for a few days more.

Cadzow told me about a native who brought in three marten skins to sell. Cadzow knew they were three nice martens, and he offered the native twenty dollars for each skin. The native said, "No, too small. I want fifteen dollars for each skin." Cadzow said he thought for a while, then said, "All right, I will give you fifteen dollars each." The native left, highly pleased with the forty-five dollars.

In the few extra days we stayed we were going to haul our stuff to the top of the summit—three steep miles up from the river above Rampart—north of the Porcupine River. A native family at Rampart House wanted us to go with them out to the flats. They told us we would easily see the flats in three days. But we had already arranged to go with the partners. They said they knew the way, as they had been across to the flats a couple of times before. I asked Hank how many nights' dog feed I should take along.

"Oh," he said, "we should be out to the flats in three days, don't take over three nights' dried fish with you."

We took along seven nights' dog feed, allowing plenty to take care of any unforeseen circumstances. When we got to the flats we expected to feed 'rats to the dogs. We were traveling now above timber line—white mountains were all around us. I noticed we were traveling in an easterly direction instead of north. By nightfall we hadn't made much headway. We were all tired, so we pulled down to a few spruce trees we could see from above. We put the night in there. Next morning both Hank and Henry were totally lost. After leaving the summit above Rampart House we never double-tripped it any more. The next morning we pulled back up to the white ridges. Again Fannie and I were the two biggest dogs in our teams. Helping them up the steep mountain ridges was a job.

After we got on top again Henry told us to wait for him until he got back; he wanted to do some scouting around for the right pass to get through the mountains. We waited hour after hour on top of the wind- and snow-swept ridge for Henry's return. Presently we spotted Henry about five miles away standing on top of another white ridge. He looked as big as one's little finger standing there. There was nothing for us to do but wait for his return. When he got back he said the way must be to the right of us, which would take us farther to the east, rather than north. We wrapped all the dog chains we had around the toboggans to help check their speed going back down the mountains. Sometimes the chains would break on going over rocks; at such times the toboggans would almost get away from us because of their speed going down the slopes.

We camped in a gully the second night. We found some dry sticks under the snow to make a campfire to keep us warm, and set up our tent. The youngsters had to be kept warm while changing their flannels after they had been tied down all day in a toboggan. Fannie used to have to melt snow for water to wash diapers every night and hang them up above the stove in the tent to dry. Joe, the youngest child, nine months old, still wore them.

The next morning we pulled up a different ridge and down the other side. We were in very rough country. It was cut up with deep ravines; some were twenty- to thirty-feet deep with straight up-and-down mud banks. Some we had to bridge in order to cross. Loose snow hanging to buck brush fell in the children's faces when it was disturbed. We had to dress Clifton's hand every day; it seemed to be slow in healing, and gave us lots of worry. After the sixth morning I asked Henry where the Crow Flats were.

CHAPTER VI — The Crow Flats

He said, "Right over there," pointing to Old Crow Mountain, the largest mountain in the vicinity. Back of Old Crow Village on the Crow River side there were many big grass meadows. Hank told me those meadows were the Crow Flats. It didn't look right to me. The Crow Flats were supposed to be straight north of Rampart House. They were hopelessly lost but they wouldn't admit it. We ran out of dog feed and started digging into our groceries, such as rice, flour, lard—this had to be cooked for the dogs in a can on an open fire. What little we could spare to the dogs barely kept them alive. They were getting so weak they could hardly drag the toboggans. We could only spare a limited amount of our grub; we didn't want to run too short ourselves with a long spring ahead of us—this was only the last of April.

Every day Hank would say, "We will be at the Crow Flats tomorrow." We couldn't turn back now; our only hope was to try to keep going ahead and look for an Indian toboggan trail or other human signs. We were getting so desperate for dog feed that we were compelled to kill two of our dogs and feed them to the remaining six. We rationed the dog meat out to make it last, so each dog only got his own small chunk. The partners were out of dog feed; they used up almost all their flour and lard cooking it up into large pancakes. We had been heading towards Old Crow village the past few days—all at once we came across a fresh dog team trail coming from the vicinity of Crow Mountain.

Old Hank said, "Well, we will turn around now. Here is the Crow Flats trail used by the Indians from Crow Village to reach the Flats."

We all turned around and headed north. Even the dogs all seemed happy that we had hit a well-traveled trail. Hank had been leading us back to Crow Village.

I said to Hank, "I thought you said the Flats were that way," pointing back to Crow Mountain, east of us now.

"Oh, no, no. We go dis way now."

Up to the day they died the partners wouldn't admit that they had been lost. Hank died at Campbell River, up on the hillside trying to gather some wood to keep his cabin warm. The wolverines supposedly ate Hank's body, thus ending the career of a colorful oldtimer, a fixture of the far north.

After following the fresh trail for about three miles we came across some Indians camping on the side of the trail. We were sure glad to see them. We knew them all from Old Crow Village. Moses Peter was there also; he was Chief of the Crow Indians. They were on their way to the Crow Lakes. I bought a good dog from the Chief, paying him fifty dollars for it. I bought two hundred pounds of moose meat from the Chief also—mainly for the dogs to eat. We rested a couple of hours and had lunch with the natives. I gave the dogs each small pieces of the moose meat to eat while they were resting—this made me think of the two dogs we had had to kill, but it had saved the lives of the dogs we still had left. I didn't dare let the dogs have all they wanted to eat—it might have made them sick. We bought all the flour, sugar, rice and lard the natives could spare. It was two easy days from there to the Flats. We went on a mile or so and camped for the night. No use rushing now. As mentioned before, the Flats are just a maze of lakes of all sizes–some ten miles in extent. After feeding so much of our grub to the dogs and buying all the natives had to spare, we were still going to be short. For this reason we knew we would have to live principally on 'rats all spring. We had plenty of Borden's canned milk for Baby Joe. He almost died once on the trip. The older boy, Clifton, seemed to have pulled through fine. His hand almost healed on the trip. The weather was still cold. The season up there is about two weeks later than Fort Yukon weather.

After reaching the Flats we went about seven miles through the lakes and set up camp. The partners went by themselves and we never saw them again until we got back to Fort Yukon. They made a small catch of 'rats. It was hard to find 'rat houses, the snow was so deep and hard no 'rat houses showed above the snow. Once in a while we found some by following fox tracks. They could smell the 'rat houses through the snow, and dig them out. I would dig down with my shovel where the fox had dug, only to find the 'rat house frozen up. I

Aerial view of the Old Crow Flats in the Yukon Territory, Canada, where James and Fannie traveled by dogsled and toboggan in 1920 to trap 'rats. This maze of lakes and streams is beloved by muskrats, and was a favorite Indian trapping area. (Photo by Richard Harrington)

thought maybe one of my dogs could do the same as the foxes, so I tied a string around the dog's neck and led him around the lake. I gave him plenty of line so he could run around freely. It wasn't long before he was smelling something under the snow. I had my shovel with me and wasted no time digging down where the dog was smelling; an open 'rat house! From the pack on my back I took out a trap and set it. The dog found eight 'rat houses for me in about two hours. I set eight traps.

The next morning I caught five 'rats—fresh meat in camp and five 'rat skins. We fed these first 'rats to the dogs; they needed them more than we did. Our first thaw out there was May 20. Many 'rat houses were beginning to show up all over the lakes. The dogs were getting fat now, and we had plenty of 'rats to eat ourselves.

We had a surprise one day when we ran across our old friend Curtis. He had run into some Indians and they told him where he could find us. I didn't think Curtis was within a hundred miles of us. He looked like a man who had dropped from Mars. As mentioned before, he had a trap line at the Howling Dog Canyon. The Indians were scared of him. They had never met him before—he wore a dirty old parka with the hood hanging over his head; the balance of his parka hung in tatters. He saw these Indians some distance off. He was so glad to see someone he waved his arms up and down to attract their attention. They thought he was a scarecrow or bushman. He yelled at them at the top of his voice. Curtis told me one of the old Indians started to pull his 30-30 rifle from the lashing of his toboggan to shoot at him.

He had been in Fort Yukon since 1897. We knew Curtis was a fine fellow to get along with, and a good entertainer, so we agreed to put the spring in together 'ratting. We all moved twenty miles farther north, which proved to be a better location for 'rats. By this time the sun never set below the horizon. Curtis had his own tent and stove. In the grub line, Curtis traveled light. He lived off the country mostly. When he arrived at the Flats he had 15 lbs. of flour; 5 lbs. of sugar; 1 can of lard; 1 lb. of coffee; 2 lbs. of butter; 6 cans of milk; $1/2$ lb. of tea, and 1 lb. of salt. He gave us his butter, coffee, and milk. We didn't want to take this from him, but he insisted we take it and said 'rats and flour were all he wanted to eat. He wouldn't accept anything for it. We used to have him eat with us quite often. It didn't take much to satisfy Curtis.

CHAPTER VI — The Crow Flats

I used to visit Curtis often to hear him talk. Our tents were close together. One day I was in his tent when he cooked a batch of 'rats. He dressed out ten 'rats and stuck them in a five-gallon coal-oil can. This coal-oil can and a couple of lard cans were all the cooking dishes he had. He put the 'rats in this five-gallon can, heads down, with their tails hanging over the sides of the can about four inches. He added water and let them boil for one hour. Then he reversed them—tails down, heads up. This gave the tails a chance to cook. The tails of the 'rats are very good eating when boiled or cooked over a campfire. Eight tails, with a small piece of bannock, makes a feed for a person. After cooking them another half hour he would thicken the juice with a cup of flour, making a thick white gravy. This concoction would last Curtis for three days. Whenever he wanted to eat a meal he would fork out of the can a whole 'rat, together with some white gravy—this didn't allow much variety, but Curtis seemed to relish it. 'Rats for breakfast, dinner, and supper. When a person is hungry and a long way from a store anything tastes good, no matter how many times one eats the same thing. Behind Curtis' sheet-iron stove in his tent he had a pile of 'rat bones as big as a large 'rat house. These were the bones from all the 'rats he had eaten recently.

About the time the snow was all melted we moved our camp to Potato Creek where it entered the Old Crow River. We had one bad storm that lasted for three days. The wind was so strong that it blew snow through our tent. It was a real arctic blizzard, impossible for any human being to face. We missed looking at our 'rat traps for three days. A few days after the storm abated, it turned warm for a while. Ducks and geese started coming into the Flats; their quacking and honking and cackling was music to our ears. We knew that spring had arrived. As the Crow Flats were summer nesting grounds, we didn't have to eat any more 'rats. From now on we saw more ducks and geese than we could have imagined. Every variety of duck came to the Flats to nest. We ate so many duck eggs that we got tired of them. Swans were seen at different times also. The Flats were alive with all kinds of small birds as well as ducks and geese. All seemed unafraid of us; they were not bothered by man and his shotgun. It seemed to us that the whole Flats turned to life overnight—there were ptarmigan in flocks by the hundreds. Sometimes we couldn't sleep for their cackling noise which started at three or four o'clock in the morning. The ptarmigan were living on low-bush cranberries that the melting snows exposed. We picked a few cans of the low bush cranberries for jam. These berries are just as good in the spring as they are in the fall when the winter snows bury them under a blanket of deep snow. We picked quite a few of the berries and made some jam, which was delicious. We couldn't make all the jam we wanted because of the small amount of sugar we had left. Cranberries are too sour to eat without lots of sugar.

It was then the first week in June. We were shooting 'rats now instead of trapping them. We hunted together so as not to get separated from one another. The lakes looked so much alike in any direction one might turn, it would have been easy to get lost if we couldn't have checked the sun. Quite a lot of water had accumulated along the south banks of the lakes. The 'rats liked to swim around in this open water and sit along the lake shore sunning themselves in the early afternoon sun. The north banks of the lakes still remained frozen solid. 'Rats sitting along these open waterbanks made easy targets for our .22 rifles. Curtis had a poor assortment of guns, consisting of a single-shot .22 Winchester pump. When we saw a 'rat I always let Curtis have the first shot because of his poor gun. If he missed, then it was my turn to shoot. Curtis carried along his 10-gauge and when-

CHAPTER VI

The Crow Flats

ever we came across two or three 'rats sitting together Curtis brought the 10-gauge into action. He would kill the three 'rats with one shot from his cannon. He turned their hides into salt shakers. We always took a lunch consisting of fried or boiled duck or goose, with a piece of thick hot cake and a butter can to make tea in.

The mosquitoes were getting so thick it was almost impossible to stand them. They swarmed in clouds. We wore head nets. They landed on our head net so thick we couldn't see through the net. Their hum almost drove us insane. They covered a .22 barrel so thick we couldn't see our sights. You could shake them off by shaking your .22 but they would land again along your gun barrel before you had time to aim. They nearly ate the two kids alive. The dogs suffered terribly—their noses and eyes were raw. We rubbed 'rat oil on the raw places, which seemed to help a little. We kept smudges around the dogs and ourselves. We had bed nets to sleep under; even at that, lots of mosquitoes used to get in the bed nets somehow. There was no fly dope in those years; at least, we never heard of any, such as Buhach, bug bombs, etc., as there is on the market today. The Crow River broke up June 1. The ice in the lakes opened up about two weeks later. One could truthfully say that the mosquitoes drove us off the Crow Flats.

The mouth of Potato Creek where we were camped had some suitable poles to make a frame for our scow. This only took us a couple of days to complete. I don't recall the exact dimensions of the scow, but it was large enough to hold seven dogs, sixteen hundred 'ratskins, bedding, two tents, two sheet-iron stoves, some cooking tools, and five people. We even brought our big toboggan down. This was put aboard the scow first, as it rested lengthwise on the bottom. Then we loaded it until it held nearly all our gear except the dogs; four dogs were placed aft, and three in the bow of the scow. Curtis had a couple of gunny sacks of dried 'rats also; he hadn't eaten enough 'rats during the past spring, so he was taking these dry ones down to Fort Yukon to eat during the summer. The scow had a dunnage strip the full length; this kept everything dry on the bottom. We fashioned a pair of oars, a paddle, and two poles. We had to live off the country going down the Crow River as we had no white-man's grub left, such as flour, sugar, and so forth. The river was full of fish; and we also had a piece of fish net about six feet long. Anytime we wanted fish to eat for a change we would stop half an hour at most, any place along the Crow River, and fish. There was no current. With the slightest breeze we would blow back upstream again. We would shove the net out from the shore with a pole—it seemed that in no time the floats would be bobbing up and down which, of course, meant fish getting caught in the net. When we wanted to take the fish out of the net we just pulled the net ashore. Our catch would be seventy-five per cent suckers. We never ate these suckers because they had too many bones. We always caught whitefish enough for ourselves, and the dogs ate the suckers. One day we didn't know what we were going to eat for lunch, when two geese flew high above us.

I said to Curtis, "Watch me knock that lead goose down with my 30-40 rifle." I hit the rear one and shot its head off.

"Gosh!" said Curtis, What a shot!"

"I do that every time, Curtis."

"The only thing is, you didn't hit the one you aimed at. You knocked the rear goose down instead. You didn't hit the one you aimed at."

His remark knocked some of the wind out of my sails. I suppose I've shot at geese a hundred or more times with a .22 rifle and larger calibers while they were in flight, but this was the first one I ever knocked out of the sky—a fluke shot of course.

Fannie started to pick the goose right away. When she had it fully plucked we went ashore and built a fire and

CHAPTER VI — The Crow Flats

boiled it—making a fire out of some dry wood we had in the scow for that purpose. There were no gravel bars or driftwood along this section of the Crow River. When one stepped ashore he would sink to his knees in mud or silt. After the goose was cooked we set the kettle containing the goose aboard the scow and shoved out on the river to let it drift as we ate. We saved the juice that the goose was cooked in to make some kind of a soup for supper. We were never stuck for eats as long as we had Curtis' dried 'rats in reserve.

We didn't know how far we were up the Crow River; however, it took us eight days rowing from where we built the scow to the mouth of the river. All day long we would be traveling around and around the sun. The two-hundred-foot mud- and glacier-banks were always sloughing off from the heat of the sun on them. In one place we saw part of a mastodon tusk sticking part way out from one of these muck-banks. We didn't try to take it. It had no value at that time. About four feet was sticking out, possibly another four feet was out of sight frozen in the muck.

One day about noon we saw smoke and, thinking it was an Indian camp, we stopped, and found an easy way to climb the high bank. We found it was our own campfire we had left early that morning. So we actually went forward about three hundred yards in one forenoon. About the seventh day we came across some Indians fishing for whitefish with nets, then drying them. The place they were fishing was called Shafer Creek, a sizable creek emptying into the Crow River. One white man and his native wife were camped with the Indians. His name was Abner (he was an old landmark from Fort Yukon). They gave us a good feed of fish and told us that it was still twenty miles to the mouth of the Crow. The country was changing now: rolling hills, gravel-bars, and swift water. Five miles below their camp the river entered a canyon. This canyon was full of big boulders and sharp slide-rock—very swift water, too. The Indians had told us to keep on certain sides of the biggest boulders. The sharp slide-rock was what worried me. If the scow scraped on the bottom, it would have ripped the canvas bottom wide open. This didn't seem to bother Curtis. I was scared, not for myself, but for Fannie and the kids. Fortunately, we went through without hitting anything. We shipped a little water from the big waves. After getting through the worst of the rapids we tied up for the night. This was the first good camping place we had since leaving Potato Creek, where we had built the scow. Gravel-bars, lots of dry wood, everything. Abner gave us a little rice to make a goose soup with. The natives gave us some whitefish. We boiled these for breakfast. I, personally, hate boiled fish of any kind, but we had no lard left to fry them in.

We reached Crow Village early the next day. Everybody in the village was looking down at us and we, in turn, were looking up at them. A fur-buyer had been at Crow Village all spring. He had already bought all the 'rats at Crow for three dollars each, including fifteen thousand dollars' worth the trader there had collected. An old white trapper who owned a motor boat took me to one side and told me to sell my 'rats quickly, as the bottom had dropped out of the fur market. 'Rates had dropped from three dollars to a dollar and a quarter per skin. This man had made a trip to Fort Yukon and right back again, traveling twenty-four hours a day with two extra pilots. On their way back they left the police and fur-buyer's mail at Rampart House. The oldtimer's name was Scotty. He tipped everybody off except the fur-buyer, of course, to sell all their furs to the buyer. So the buyer bought every 'rat at Crow for $3.00 each.

We tried to sell him our skins but he had already spent all his money. He asked us to wait until he got down to Fort Yukon. He said he expected money to be waiting there for him from his company. I knew what that meant.

CHAPTER VI — The Crow Flats

So we had to take our 'rats to Fort Yukon with us. We had been a little too late getting to Crow Village. We were all so busy and excited that we forgot we were hungry. The store owner was leaving for Fort Yukon in an hour or so. I asked him if we could ride down with him, and how much he would charge us. "You can ride down with us. The charge will be thirty dollars. Bring your scow alongside my barge and put your junk aboard." There was lots of room, as his barge could hold ten or twelve tons. The distance to Fort Yukon was three hundred miles. I ran up to the trader's store and bought a box of groceries, including canned fruit. The trader always left someone in charge of his store while away on trips. We would make it to Fort Yukon in two days and nights.

On our way down the Porcupine River we stopped at Rampart House to let the police off and to pick up the fur-buyer's mail. I noticed the fur-buyer's mail consisted mostly of telegrams. After he opened part of his mail he threw up both hands and laughed. He knew he had been jobbed. But he was a good sport. He was told by his boss not to buy any 'rats or anything in the fur line and to come right back to Fort Yukon. He left sixty thousand dollars in Old Crow Village. He gave the trader a letter of credit for forty thousand dollars; the company the fur-buyer was buying for tried to back out of paying this to the trader. They couldn't make it stick. When the buyer reached Fort Yukon he was fired by telegram. The 'rats he bought at three dollars each just about broke the company he was buying for.

We sold our 'rats to a Fort Yukon trader for a dollar fifty each. We had over a thousand of them. The trader held these 'rats for more than a year, and finally had to sell them at seventy-five cents each. Had we stayed down at Fort Yukon we would have caught just as many 'rats and sold them for a better price. With all the hardships and danger we went through we hoped we would never see the Crow Flats again. Curtis had seven hundred 'rats and two gunny sacks of dry 'rat meat. Curtis held his 'rats for quite a while; he liked to speculate in furs.

We didn't see Hank and Henry until they showed up at Fort Yukon about the middle of July. They had over four hundred 'rats. They still wouldn't admit that they had been lost trying to find the Crow Flats.

CHAPTER VII
Back at Fort Yukon

We were sure we had had enough of the Crow River Flats. After reaching Fort Yukon we rested up for about two weeks, regaining lost weight. When we had left Fort Yukon in early April for Crow Flats I weighed one hundred and sixty-five pounds, my weight now was down to a hundred and thirty-eight. Fannie's weight was down several pounds; even the kids looked sick. We were lucky we brought them both back alive. We were going to stay in Fort Yukon until early fall and just take things easy. I overhauled our one-lung gas engine, as we planned to go back to our old trapping ground up the Salmon River to do some more trapping. Ordinarily if one came in town too early in the spring one would have a hard time trying to find something to eat—with the stores all sold out of grub. Trappers were in the habit of bringing enough grub back into town with them to live on until the first boat arrived from White Horse; this was generally around the first week in June, or after the ice cleared out of Lake Laberge so the boats could get through. Curtis always brought something to eat in with him even if it was dried 'rats. There was some king salmon in the Yukon at this early date but one couldn't eat salmon straight—they are too oily.

An old Russian commonly called "Scow" Davis used to drift down from Dawson once in a while. He peddled eggs, bacon, apples, oranges, canned fruits, potatoes. He peddled from Dawson to the mouth of the Yukon River. The last time he stopped there his scow was swamped when a heavy wind suddenly sprang up. There was no time to move it to a more sheltered place. The high waves washed everything out of his scow in a few minutes. There was no chance to save anything—his total cargo was a loss; it floated down the Yukon River. Every boat in town struck out down the river to try to salvage what they could in spite of the high waves. The wind was blowing inshore and lots of stuff was blown to the beach. The old man, however, never got anything back. Everything salvaged was hidden in the brush until he left town. Almost everybody had more bacon and eggs than they ever had in their lives before. The wet eggs would not keep long. Scow's canned stuff all sank to the bottom. The bacon that was salvaged wasn't hurt at all. By taking the wrappers off and drying it out, it was kept from molding. Some of the eggs were picked up six miles down river.

It was a tough break for the old man—he was angry at everybody. He called them all robbers and thieves for not returning everything that was salvaged or picked up. The old man didn't remain long in Fort Yukon. He pulled out with an empty scow. He swore that he would have every one of us behind bars as soon as he got to where there was some law. Nobody heard from old Davis again until the next spring when he came floating down the Yukon a day or so ahead of the first steamboat. He had his usual load. He tied up his scow in a more sheltered spot. He never mentioned anything about the ordeal he had gone through the previous spring. Everybody patronized him, more or less. We all liked to see the old man come down. It would have been better if he could have arrived ten days sooner. He could have sold all his cargo right there, when everybody was hungry and the stores all sold out. He used to buy all the bearskins in town that were offered to him. I sold him a large grizzly bearskin for a suit of clothes, about a ten-dollar grade. I had no use for the bearskin and much less for the suit of clothes. I sold the suit to a native afterwards. This was to

CHAPTER VII   Back at Fort Yukon

be the old man's last journey down river. Somewhere along the way between Fort Yukon and Saint Michael he was ambushed and killed by a shot from the shore. His murderer was never apprehended. He was robbed of what money he had, which must have amounted to quite a sum. His scow was found tied to shore with his bullet-riddled body lying on the bottom.

By August 20, we were ready to start upstream again to our old trapping grounds. We spent six easy days going up to our camp with a heavy load: about one and a half tons, including seven dogs. The little motor never faltered

From right to left, five Carroll children, Joe, Clifton, Annie, Mabel, and Tommy who was born in 1927. Taken in the early 1930s. (Courtesy Carroll family)

once. We were going to make this our last year of trapping. We had been laying away a few dollars each year. We planned to build a small trading post at Fort Yukon the next year. We had three small children now. Two boys and one girl. This was too many children to be hauling around through the brush, "nigger-heads" and swamps.

We had two toboggans. Fannie's was a medium-size toboggan. We spread the bedding full length and laid the kids on top. Clifton rode behind sitting up, while the two young ones laid side by side, well-covered, and lashed to the basket of the toboggan with a small piece of rope.

125

This lashing used to be Fannie's job every morning when we were on the trail away from home. In my toboggan I hauled the tent, stove, dog feed, grub, and traps.

After we unloaded the boat and put everything in the first cache we were to make a trip up the river about fifty miles to cache some dry salmon for dog feed that fall. This would save us from having to haul it up later on with dogs. Old Deagle had left the country. We intended to use his cabin and trap his side lines. We made the round trip in four days. We saw no game, but lots of big bear tracks. We had bought, from some trappers' supply catalogue, two large bear traps. To set a bear trap we made a V-shaped pen from small spruce trees cut in six-feet lengths. We set the bear trap in front of this pen and threw some bait in the back of the pen. Meat is preferable for bait because they have too much fish to eat; however, a piece of rich salmon will do. We had one trap set up river and another trap down river about one mile apart. To find out if we had a bear, without visiting the traps every day I would get up and step outdoors just before daybreak and listen for half an hour. If we had a bear caught, he would always bellow like a domestic bull, and sometimes it would be a long drawn-out bawl. Later on we would go to the captured bear and shoot him.

There weren't many moose on the Salmon River in those days, but we always managed to get a couple each year. The dogs liked bear meat better than the salmon. It had to be cooked for them. After freeze-up they would eat it frozen. The bears were rolling in fat, and so were the dogs after eating the bears. We ate a few lynx ourselves. We would pick out the fat-looking ones that had been recently caught. After they stay in traps too long they are no good to eat. The lynx meat is light in color, like pork. After being cooked, the meat tastes good—when you have no other meat to eat. The thought of eating a cat is all that bothers you. We used to pot roast what we ate; when we had moose meat to eat we passed up the cats.

CHAPTER VII

Back at Fort Yukon

Early one morning, as I was listening outdoors, I could hear a bear in one of our traps. This bear was making more noise than usual. As soon as the bear saw us approach him he charged us so hard that he threw himself over backwards. He blew steam from his mouth and roared. The cold, frosty morning must have caused his breath to turn to fog. We shot this bear and skinned him. His fur was that gunnysack color. He must have had four inches of fat on his hips; his weight was about five hundred pounds—a big bear for that part of Alaska. He had chewed big chunks out of all the trees he could reach. A bear caught in a trap can hear you coming quite a long distance and he never makes a sound until you come close to him.

I had caught another at this same place. We could hear his long, drawn-out bawl. I went down to where the bear was caught alone in my canoe. When I got to the place the bear should be I couldn't hear any noise. I climbed the bank without making a sound. I couldn't see or hear anything at first; then I saw where something had been dragged across a little slough and up into the brush. I went across this slough below where the bear had gone in. I stopped to listen two or three times. I still couldn't hear a sound. I was sure the bear was close at hand. I carried a shotgun and a rifle with me, I figured if the bear charged me I would use the shotgun, but if I saw him at a distance I would use the rifle. All at once the bear reared up on his hind legs and charged at me with his mouth wide open. I was so surprised and scared that I turned and ran—still holding the two guns. I jumped into the canoe and paddled across the slough—all this time the bear was fighting to get loose, but he was so tangled up in the brush he couldn't pull himself free from the trap. I climbed up the bank of the slough and shot him dead. Never carry a gun in both hands when you are trying to locate a bear in a trap. If I had only one gun I could have shot the bear when he reared up at me. I never reset that trap again.

The Salmon River was late freezing up that fall. We each had a waterproof tarpaulin. In loading up our toboggans we would lay these tarpaulins full length and load our stuff in on top. By doing this, we could cross any open water eighteen inches deep without getting anything wet. We came to a place where the whole river was open and we had to cross this water somehow in order to hit our trail on the other side. The water wasn't very deep, but it was quite wide. We had a pair of old hip boots that leaked a little along. These were brought along because we knew we would run into lots of open water. It was early in November; we had no cold weather yet. I changed my leather-top shoepacks and slipped on my hip boots; then I waded the river to test the depth of the water. One place I waded across the water, it was up to my knees but it wasn't very swift.

I waded back again and we decided to take the kids across first. I held the handle bars. We had good dogs those days, you could drive them as one would drive a horse. We crossed without any trouble. I unharnessed the dogs and tied them to a snag, just in case they might spot some game and run away with the kids, which were lashed in the toboggan. I waded back across and carried their mother over. My toboggan was so large I got a pole and sat astride my load. It floated like a canoe. The pole helped to keep the toboggan on an even keel. I crossed a little below where I had crossed with the kids. The dogs had to swim part of the way across; at that, we made the crossing without getting anything wet. The oiled tarps had kept everything dry.

Before we left home I had set a bear trap a short distance below our cabin; while we were away a bear got caught in this trap and he made away with the trap and toggle. He left no trace of which way he went. The toggle, a small one, was six feet long, six inches in diameter. He could have chewed this toggle to slivers in no time. I figured the bear got tangled up in thick brush and starved

CHAPTER VII — Back at Fort Yukon

to death; so the bear with the trap was forgotten until I made a trip to Fort Yukon.

Just before the holidays everyone was talking about a brown bear that was killed on Christian River; this would be sixty miles northwest from where my bear was caught. They told about the odd track they saw; it looked like the track of a bear dragging a log to one side. This frightened the Indians; they had never seen anything like it in their lives. They went back to Venetie and got some more hunters to track the queer animal down—Venetie is a small native village. On their way back they ran face to face with the bear, which was traveling on the trail the Indians had made the day before. As soon as the bear saw them he blew hard and stood up on his hind legs. The trap was on his front foot; the Indians said he held the toggle and trap under his arm as a human would. They also said the toggle was iced-up from his dragging it through open water and snow for so long. They said it must have weighed a hundred pounds with all the ice it had gathered. They shot the bear, of course. He was so poor that the dogs wouldn't eat the meat. So he was left lying where he fell.

The trap that he had been caught in was a twenty-pounder. One wonders how the bear traveled so far over all kinds of rough country, brush, timber, and windfalls, and "nigger-heads," with such a handicap. He probably carried the toggle and trap under his front foreleg as the natives saw him do. He must have lived on mice and ground squirrels, or perhaps he found a dead bull moose that got killed in a fight with another bull moose. The Indians who shot the bear kept the trap—they took it back to Arctic Village, where it remains to this day. It is unlawful to set these bear traps today.

We were making another trip up the river to the place where we had cached the fish; about ten miles this side of where we left the fish we ran into caribou. They were all traveling downstream. When the dogs spotted the caribou they went wild to chase them. Just before this we were sorry for the dogs because they all appeared so tired from pulling the loaded toboggans, but after they spotted the caribou we had a hard time to keep them from running away. We both jumped on our loads and away the dogs went, making, it seemed, about ten miles an hour. They kept this speed up for over an hour. There were now caribou behind us, ahead of us, and on both sides of us. The herds were not massed, but scattered. We were in danger of being stampeded. We made the cabin where we cached the fish in record time. This run of caribou kept up for several days. After that, the herds thinned out, but they kept coming, a few at a time, for another week or so.

We hadn't shot any caribou up to then. When we arrived at the fish cabin the caribou had a trail up the riverbank leading to the door of the cabin, which was about twelve feet back from the bank. When they hit the cabin it split the herds in half momentarily. I am quite sure had the door of the cabin been left open the caribou would have run inside the cabin. Every direction you looked you could see caribou. Finally the caribou were so familiar to the dogs that they didn't bark at them any more. I shot six of the animals; this was all the caribou we wanted. Our dry fish looked intact; nothing had bothered it. For variety, we fed most of the caribou meat to the dogs. We felled some trees across the trail leading up to the cabin. We could now step outside without stepping on a caribou or being knocked down by one.

There was a large mink camped in the cabin under the fish pile, living off the dry fish. He had a hole in the back corner of the cabin where he used to come in and out of the cabin. The mink was in the cabin right when we arrived. We could hear him under the fish pile. We hadn't unloaded our toboggans yet. We shoved them in the cabin loaded, with the kids still lashed on. We tied our dogs back of the cabin where there were some very old

CHAPTER VII

doghouses. We had some H-3 Jump Traps in t
We took one of them back of the cabin where
hole was. I set the trap and told Fannie to hol
over the hole while I went inside with a stick tc
mink out. The mink charged out full speed, knocking Fannie and the trap over backwards. There was a spruce tree right close to the corner of the cabin. The mink ran up this tree like a marten. That was the first time I ever saw a mink climb like that. It climbed about halfway up. The tree had lots of green branches which partly hid the mink. I had a .32 belt gun.

I said to Fannie, "I know I can't hit anything with this pistol because I never did before." It was the only small shooting iron we had outside of my 30-40 which would have blown the mink to pieces. So I pointed the pistol where the mink should be and pulled the trigger; down came the mink, dead—shot through the neck. It was the first thing I ever killed with the pistol. Heretofore, I carried the gun around only for the noise it made when fired to scare bears. The mink added another pelt to our collection of furs.

We stayed at the cabin a few days, doing nothing. The caribou had thinned out considerably. I made a short trip about five miles back from the river in search of fur signs, but the caribou had the whole country beaten down with their tracks. We waited a short time longer hoping it would snow so we could see fresh fur signs. We shot five more caribou just in case we might need the meat to eat ourselves or to feed to the dogs. We expected the run to stop anytime. We hadn't had a snowfall for a long time so we decided to go back down to our home cabin. We stopped awhile at Deagle's cabin. The caribou didn't come down the Salmon River this far, but turned off in an easterly direction towards the Porcupine River country.

At Deagle's cabin we set some traps on the two short trails which he had cut east and west of his cabin. From there, we went on to our home cabin. The river was frozen

131

over solid. We didn't have to ford the crossings any more. After we arrived home I looked at some side lines, close in, that I had set out before we went up the river. Fannie and the kids stayed at home taking a well-earned rest.

I often wondered in later years how those young children pulled through all they went through alive. They were dragged all over the north country in all kinds of weather with temperatures ranging down to sixty below zero at times, lashed on a toboggan in winter, or a boat in summer. But they didn't seem to mind it as long as they had a change of flannel once in a while and a bottle of Eagle Brand milk. As far as we ourselves were concerned it was all in the day's work.

From left to right, Annie, Mabel, Tommy, with their mother, Fannie. Taken in 1930. (Courtesy Carroll family)

One night, I was coming home from a round trip down river; on my way back I had to make a crossing of the river, over a trail I hadn't been on for a couple of weeks. About halfway across my lead dog stopped and was smelling the trail. I hollered "Mush," to him, which means in dog language "Go!" But he wouldn't budge an inch forward; he kept jumping from left to right trying to turn the swing dogs around. I hollered "Whoa!" and got my axe from the back of the toboggan. I walked up to the lead dog to find out what was wrong. I swung the

CHAPTER VII Back at Fort Yukon

axe with one hand, chopping into the ice, after I got to the lead dog—the axe slipped out of my hand and sank to the bottom. There was nothing but snow holding the trail up. I got back to the handle bars of the toboggan and called the leader around. I beat it back to the shore. It was quite dark and I had a hard time trying to find a pole to test out another crossing. The water at this place was about ten feet deep—dead water. I don't know how I escaped from falling in. I finally found a suitable place to make a safe crossing.

The Salmon was a dangerous river to travel on. If the weather was forty below it would freeze up safe enough for travel; when the weather moderated to about ten below the ice would melt, leaving only the snow holding in many places. The river was full of warm springs, which are very dangerous for night travel. In the daytime these melted the snow and showed up plainly. I had lots of close calls but I never did fall through the ice over my head. On many occasions we have seen where moose have drowned in the Salmon River by falling through the ice and being unable to get out. Their heads and the hump on their shoulders was all that showed above water.

We intended to make one more trip up to the fish cabin. The days were getting longer now by several minutes each day. During the shortest days the sun never got above the horizon. We were over a hundred miles inside the Arctic Circle—at that, on clear days, we had four hours of daylight, and on cloudy or stormy days, about two-and-a-half hours of dim daylight. We had no trouble going up the river this trip. Very few fresh caribou tracks were seen. It was too late to set out any new trap lines. We stayed there a short time to give the dogs a chance to eat up the surplus dog feed; we didn't want to drag any of it down the river, except enough to last us on the trip home.

After we got home we picked up all our traps that were still set out; then we just killed time until the ice in the river broke up and floated downstream. No use to overhaul our one-lung motor—it had been in good running order last fall when we put it away for the winter. Our collection of furs was average. That season Johnny Olson, a fellow living at Eagle, Alaska, had trapped far above us in the Endicott Range foothills. He trapped mostly for foxes, he had a nice collection—mostly crosses. We invited him to stay with us for the ice breakup and go down to Fort Yukon with us in our boat. We had plenty of extra fish to feed his dogs. Johnny was the only human being we had seen in many months. After we got down to Fort Yukon we planned to turn our trap line over to Olson.

Before reaching the mouth of the Salmon we stopped at a nice gravel-bar to brew some coffee. After lunch we were going down to look at the Porcupine River to find out if the ice had run out yet. Sometimes we used to get to Fort Yukon ahead of the Porcupine ice—this was dangerous practice. Should we run into an ice jam in the Porcupine River below the Salmon and have the Porcupine ice come down on us, we would be in a bad spot. After we finished our lunch and were about to pull out again we heard a noise that sounded like a freight train coming. We knew this was the Porcupine River ice breakup—a spectacular sight—a wall of water and ice twenty-feet high rolling down to the bottom of the river. When it passed by the mouth of the Salmon it shot ice and water and our boat several miles up the Salmon River. It flooded the countryside so badly that we couldn't find a piece of dry ground to camp on. We had to live on our boat three days. The ice flowed down for seventy-two hours. After the ice passed, the river dropped several feet drawing the ice back to the Porcupine. If this hadn't happened, we might have been unable to get through to the Porcupine River for some time.

The Porcupine River at this high stage of water is very swift, traveling several miles an hour. All the low-cut banks were underwater. When we arrived at Fort Yukon we could see that the town had a bad flood. Ice was piled up everywhere on top of the Yukon's low banks. The town itself was full of ice cakes; but everyone seemed cheerful, going about the task of cleaning up their respective property. Everybody loses in a flood; nearly every cabin in town had water in it. The natives lost most of their belongings by not being there to dry their goods out. When they did come back from their 'rat camps, it was too late to save anything. Everything was moldy and rotten. Some doors were broken open and their wet, muddy mattresses were thrown out to dry. The stores were naturally the heaviest losers—mostly in flour and sugar. They were sold out of almost everything else until the first boat came down—which would be another ten days. It was lucky for them that the flood happened when it did—before their spring cargo of groceries had arrived.

CHAPTER VIII
The K-Brothers' Strange Disappearance

It was the winter of 1922 that the K-Brothers were reported missing on the head of the Old Crow River—somewhere north of the Crow Flats. Nobody seemed to know exactly where they had gone. They had two dog teams loaded down with groceries and dog feed. It was reported that they were going to Big Bear Mountain to prospect, or to the coast to trade a little. At about this same time a report came in from Arctic Village that two native youngsters had shot and killed two bushmen. One of the bushmen was tall and one was short. This sounded like a description of the K-Brothers. One of them was short and the other tall.

These so-called bushmen used to come in from the Arctic Ocean and try to steal the native womenfolk. This happened many years ago. The oldest Indians still talk about it. They actually believe that such people existed and lived in caves in the wintertime. According to the Indians, they wore bearskin suits with feet and claws left on. Dressed in this way, they could get around in the wintertime without being observed. Anybody who came across their tracks would think they were bear tracks.

The marshal's office in Fairbanks was notified of the rumor and sent a deputy to Arctic Village to investigate. The deputy's name was Scott. The distance from Fort Yukon to Arctic Village was around two hundred miles by dog team. Thirty miles of this distance was above the timber line. It was too far for one day's travel, with only about four hours' daylight, so one night's dry wood had to be hauled along for several miles. Scott had a native guide along. When they arrived at Arctic Village, he questioned the whole village. The two youngsters stuck to their original story that they shot two bushmen. The weather was foggy, but they were sure they had killed both bushmen. They were afraid to go close to the dead bushmen, so they ran home frightened to death, and told their parents what they had done. They took the marshal out on the mountainside and showed him where they had shot the two bushmen. They found nothing, of course. The marshal came back to Fort Yukon and reported what he had been told by the two boys. There the matter rested until spring breakup.

After the rivers broke up and were clear of ice a new marshal by the name of Hocks came over to Fort Yukon from Fairbanks and organized an overland search party. Hocks brought a boat and a one-lung gas engine that could hardly buck the river current. The boat was a twenty-footer with a five-foot beam. The marshal took three men with him to act as guides and searchers. Their names were Hobo Bill, the main guide, old Jim No-Sick, Marshal Hocks, and me. We had just gotten to town from our trap lines on May 23, which was an early breakup. I didn't think so much about going on such a trip after just arriving in town, but I decided to go anyway. We had a boatload of grub, gasoline, oil, two axes (in case we lost one), rifles, shotgun, ammunition, tent, stove, bedding, mosquito bars to sleep under, etc. Our first day up river from Fort Yukon we made Seventeen Mile Point of the Porcupine River. Our plans called for forty miles up the Porcupine to the mouth of the Salmon River, then up the Salmon River two hundred miles. We were to cache our boat and provisions and cut crosscountry to Arctic Village with packs on our backs.

We camped at Seventeen Mile and slept until noon the next day. I was the first one to wake up; as I fell heir to

CHAPTER VIII The K-Brothers' Strange Disappearance

the job of cooking, it was my duty to be the first one up to prepare breakfast. Naturally, the first thing I did was to look down at the boat. It was hanging almost straight up-and-down by the bow painter. We were lucky the rope was a new half-inch Manila, one which held the full weight of the boat. At this time of year those rivers may fall or rise several feet in one night. I hollered at the top of my voice for everybody to get up quick; that the boat was swamped. Everything in the back of the boat was wet, including two fifty-pound boxes of bread. A new shotgun slid off into the river and was lost. You might say half our outfit was wet. We spent a whole day there trying to dry out stuff, such as beans, rice, crackers, tea. We were lucky the sugar was in the bow of the boat and didn't get wet. For a while we thought that we might have to go back to Fort Yukon and replace the grub we lost. But we decided against this. We could make out by living partly off the country. We could kill moose and partly dry it. No-Sick Jim and Hobo would rather have dry meat to eat than white-man's grub. The boat's engine was partly under water—the magneto didn't get wet, which was lucky for us. The next day we pulled out and made it to the mouth of the Salmon River. We camped overnight and part of the next day. One of the crew shot two black ducks, real fish-eaters, with no white on their wings. They were big ducks. We had to use pliers to try to pick them, but this was too slow, so we burned their feathers off. This worked pretty well. As I was the cook, I was going to make a big mulligan—a black-duck mulligan. First I boiled the ducks for three hours in a cast-iron dutch oven—the best kettle there is for cooking over an open fire; then I added some potatoes and onions. After the whole thing boiled for another hour, I set it off the fire, lifted the lid, and looked at it. It looked very appetizing with a half-inch of red oil floating on top of the mulligan. We were all hungry. Marshal Hocks filled his plate up first. It tasted and smelled so strong of fish he couldn't go it. Even old No-Sick Jim had to turn it down. Jim could eat anything digestible. Hobo Bill said he couldn't pass up such a rich-looking stew. But Hobo ditched his plateful along with the rest. I lost my appetite just smelling it. They all said it tasted like otter. I told them I couldn't substantiate this fact because I hadn't eaten a stewed otter yet.

The gang was willing to settle for some canned beans and crackers and coffee. The hardtack was swelling up, getting bigger all the time. We had three boxes or more, although we started out with two boxes. We didn't know what to do. We had no containers to put the surplus crackers in. We decided to lay off another day and dry out the crackers, rice, and beans. To do this we spread out our big boat tarpaulin. We spread the crackers out on this canvas. It looked like a half-acre of crackers; some of them had already started to mold. We spread the beans and rice out on smaller pieces of canvas. The damp crackers collected lots of fine sand and dust that the wind whipped up from the sand bar we were camped on. We eventually had to ditch the whole works. We tried to brush the sand off the crackers with a whisk broom, but this wouldn't work. We had to eat bannock for bread from then on. Bannock is a thick dough, cooked in a big frying pan over hot coals by a camp fire.

Marshal Hock's plan was to go as far as we could up the Salmon River—possibly one hundred and seventy-five miles by his boat. The farther up the Salmon we went the swifter the water became. Finally, the water got so rough and swift that Hocks' boat could no longer buck it. Three of us tried walking along the riverbanks and bars. This lightened the boatload about four or five hundred pounds. But we still made very poor progress. Walking along the banks was almost impossible due to thick brush, windfalls, and deep sloughs cutting into the river. We tied up the boat at a good place. This was a hundred and forty miles up from the mouth of the Salmon

CHAPTER VIII — The K-Brothers' Strange Disappearance

River. We decided to walk across the mountains from there to Arctic Village, which was estimated to be sixty miles away.

After caching all the grub that was left and the boat in big trees, each of us had a forty-pound pack. Before evening they felt like hundred-pound packs. It took us six days to reach Arctic Village. On our way we saw only one small herd of caribou. We needed some meat badly. We used to fish for grayling in the numerous creeks we forded. We never were able to catch all the grayling we could eat. As soon as the caribou spotted us they started running up a side hill; we were all shooting at them, but nobody made a hit. When they reached the top of the ridge they stopped and looked down on us.

I turned to Hobo and said, "Hobo I am going to take one last-chance shot." Which I did. Old Jim No-Sick said he saw one caribou drop. Everybody was happy. We all walked up the side hill, which seemed like a mile to the top. Sure enough we had a dead caribou, shot through the shoulders. I aimed about six feet above their backs with my 30-40 Krag. We dressed the caribou out and dragged it down the hill. We started a campfire, skinned the caribou, and cut some steaks from it at once. There happened to be a small creek close by. We put the warm steaks in this ice-cold creek to chill them and get the animal heat out. Having had no meat to eat for some time, we all sure enjoyed those steaks, especially the natives who roasted a whole rib-side over the campfire. They ate all this, together with part of the fried steaks.

We were getting above timber line now; this made the walking much easier. We took as much of the caribou meat as we could pack and cached what was left on the ground, covering it up with the hide. The weather was pretty cold up there; there were still patches of snow on the ground. We had a hard time making a campfire to cook on; there was nothing but small green brush to burn, and this was wet from snow squalls and rain. Instead of our packs getting lighter they were actually getting heavier. We had no pieces of canvas to protect them from getting soaked with rain. Once it rained for twelve hours. We were soaked to the skin and had to wring our bedding out as one would wring out a dishcloth; there was no sunshine to dry anything out. Our sugar was carried in cloth sacks in our packs—it was dissolved and lost. When one is packing on his back, he can't take along everything he would like to—such as a tarpaulin to sleep under for protection from the weather. We all arrived at the village okay, but tired out. The natives treated us very well, and gave us the best they had to eat. They also gave us dry bedding and a cabin to stay in.

My bedding consisted of a cheap, small comforter filled with cotton, which never did dry out on the whole trip. The two youngsters' story hadn't changed a bit from what they had told Marshal Scott. We searched the whole area for miles around with no trace of the dead bushmen's bodies. The entire village cooperated with us in the search. So, there was nothing more we could do. We stayed at the village a whole week. Hobbs and No-Sick Jim's blankets were well dried out. Hocks' and mine were still damp. Marshal Hocks paid the natives for everything we bought which consisted of dry meat, flour, sugar, tea and lard.

An old native by the name of Simon accompanied us on the trip back. Old Simon guided us back on a different route, which he claimed was a shorter way. It took us five-and-one-half days to get back to where we had left the boat. The second day out from the village we came to a big creek we had to ford which Simon forgot to tell us about. The water was very swift and up to our waists. If we hadn't each a pole to steady ourselves our feet would have been swept from under us. It rained nearly all the way back. Our bedding was soaked again which made our packs so heavy we could hardly stagger along under them. The third day back we reached the place where we had left the caribou meat. No-Sick Jim

CHAPTER VIII   The K-Brothers' Strange Disappearance

had to steer old Simon from his short cut or we would have by passed the meat. We were just taking a chance that the meat was still there, and had not been eaten up by a bear or wolverine. The ravens had gotten ahead of both wolverines and bears, and cleaned the meat up themselves. This was a disappointment to a hungry and tired bunch such as we were.

The last twenty miles to the boat we had only one hot cake apiece. It took us eight hours to make this last twenty miles. We began to worry now about the boat. Was it where we had left it? Had something chewed the rope that was holding it? Even a rabbit could do this. A sudden rise of the river, which is possible at that time of year with the Salmon, can make it a raging river full of driftwood and trees which might hit the boat and break the line holding it. The wolverines could have climbed the trees where our grub was cached and knocked it to the ground. The bears could spoil what they didn't eat by biting holes in the canned goods.

But the boat was the same as we had left it. The river had dropped considerably in our absence, which left the boat high and dry. The first thing we did was to make a good campfire and prepare something to eat. Most of us ate so much on empty stomachs that we got sick from stomach pains. We all laid down, and we didn't wake up again until the next morning, when we felt much better. The boat had dried out so much that it was full of open cracks. We had to take everything out of it, mostly gasoline, to lighten it up so that we could turn it over in order to caulk the open seams. We had nothing but gunny sacks to do the job. These we cut in four-inch strips. This gunnysack caulking did the job. Where the boards were cracked and couldn't be caulked, we rubbed grease into the cracks. We turned the boat right side up, and shoved it in the river. We then put aboard all our junk. The engine started immediately. We made the mouth of the Salmon in fourteen hours. No-Sick Jim and Simon went to Fort Yukon from the mouth of the Salmon in a small boat No-Sick Jim had left there early in April. He dragged it up from Fort Yukon with his dogs intending to use it to shoot 'rats with when the lakes opened up in May.

Marshal Hocks had orders before leaving Fort Yukon to continue the search should no trace of the brothers be found at Arctic Village. We were to proceed to the head of the Old Crow River, which they were last reported heading for. This meant another boat trip of four or five hundred miles. We were running short of grub by this time. The gas was holding out though. We had forty gallons left here at the mouth of the Salmon. We could replenish our grub stores along the way. Healy and Strom operated a small store at Old Rampart, which we could reach in four days' travel up the Porcupine from there. When we arrived at Healy's and Strom's place they let us have whatever we wanted in the grub line, including twenty gallons of gasoline. This would last us to Old Crow Village. The engine we had was easy on the gas.

The next day we arrived at Rampart House in Canadian Territory. There were two Royal Canadian policemen stationed at Rampart House. One of them accompanied us on the Crow River trip. At Old Crow Village we bought more grub and gas. This was our last chance to buy anything. Schultz, the trader at Crow, was short of supplies himself, but he let us have what he could spare, including more gas. The first twenty miles up the Crow River is swift water so we made slow time through this stretch. I had already gone down the full length of the river that year in a loaded canvas scow. After we passed through twenty miles of swift water the river was very slack. We could go up river just as fast as coming down.

After we had gone about one hundred miles upstream from the canyon, at daybreak one morning, with a low fog hanging over the river, we spotted something in the river far ahead of us. At first we thought it was a bear. Hobo said it looked like two bears. We made ready with

CHAPTER VIII The K-Brothers' Strange Disappearance

our rifles to start shooting when the policeman, who had been looking through his field glasses, hollered, "Don't shoot! It is two men on a raft dressed in caribou-skins with fur side out, floating down on a raft."

It turned out to be the missing brothers, who nearly got shot after all. After a deluge of questions and answers about where had they spent the winter, we described the big search for them which had started early the previous winter, and told them about the two boys who had shot at them thinking they were bushmen. One of the bushmen had been tall and one was short. The brothers looked so bewildered I had to laugh in their faces. We had them abandon their raft and come aboard the boat with their gear. The brothers rode back to Fort Yukon with us.

This happened thirty years ago. The brothers seemed to have drifted apart since then. The short one turned to fur-buying for a Fort Yukon trader. He made trips to the Arctic Coast buying white fox from the Eskimos. White fox were averaging forty dollars each those days. The last trip the short brother made, he bought four hundred white fox. He couldn't haul them all on one trip; he had to double-trip them through the mountain passes. An Old Crow native making a routine trip to the coast came across the short brother dead on the trail. It appeared he just got back from the coast with his second load of foxes when he was stricken with a heart attack. He was lying partly over his campfire. The native returned to Old Crow to notify the police, who took charge of the situation. His body was taken to Old Crow for burial. The Eskimos took quite a beating. The short brother bought more than half the foxes he had "on the cuff." The Eskimos never received any compensation for the loss of their furs. The tall brother died shortly after his brother died, by eating something that didn't agree with him. I never learned what it was he ate.

CHAPTER IX
Home at Last

In the spring of 1922, we decided to kill a moose and dry the meat. We both liked dry moose meat. We could take this to Fort Yukon with us and have something to eat before the arrival of the first boat from Dawson. When we were going down the river I noticed lots of moose tracks four or five miles above our cabin. While we were picking up the last of our traps, the caribou probably scared them down from the upper part of the river. Moose are scared of caribou because they make a clicking sound with their hoofs while traveling. It is a sound the moose do not like until they get used to it. The first windy day that came along I went up the river where I had seen all the tracks. When there is a strong wind blowing the moose can't smell your scent. If he does, he can't tell where the scent is coming from. The wind whirls your scent in all directions.

I walked right on top of a moose. He saw me just as I saw him. He jumped ten feet from his bed into the thick brush. Usually, after you jump a moose, he will stand still for a split second and give you a chance to shoot once if you have your gun ready. I fired six shots at this one, all the shells I had with me. Then I didn't know what to do. I always carry my gun full of shells and put five or six loose ones in my pocket. Apparently I had forgotten to do this. I followed the tracks a short distance and noticed some blood in the snow. There, just ahead of me, was the moose lying down. On the sight of me approaching he stood up and looked at me, shaking his head. I didn't like this so I went back a short distance and stood by a tree, ready to climb it if I had to. But the moose just stood there and never moved. He must have been hit somewhere in the body. I just stood there under the tree thinking and hoping the moose would drop dead. But he refused to do so. I didn't like the idea of going home after more shells; this would mean a round trip of seven miles. So I just waited and waited. I was getting extremely cold, especially my feet, from standing in the snow in one place so long. I threw sticks at the

James and Fannie Carroll's cabin in Fort Yukon, purchased in 1922. Note the team of horses bringing in a load of firewood across the frozen Yukon River. (Courtesy Carroll family)

moose; he just turned his head and looked at me. I broke down a long dry pole—I thought possibly I could push the moose over. But nothing doing. He wavered a little and took one step forward. I thought of going home and letting the moose stand there; but this wouldn't do. I was sure the moose was bleeding internally and would drop dead in less than two hours. By morning the meat wouldn't be fit to eat.

I had another idea. If the moose allowed me to poke him with the pole, why not fasten my seven-inch hunting knife to the end of the pole and stab him? It would

CHAPTER IX — Home at Last

be a cruel thing to do. When hunting I always carry some strong rope with me. With this, I fastened my knife to one end of the pole. Before I had a chance to jab the moose he fell dead. Our dried meat for the summer was assured now. It took me about three hours to skin and dress out the moose.

It was long after dark when I got back home. Fannie was worried that something serious had happened to me—either I had fallen through the ice or accidentally shot myself. I didn't like the idea of having to walk so far in the dark carrying an empty gun on my shoulder. I hauled the meat home the next day with the dogs. It was in fine condition. In preparing the meat for drying we cut all the meat from the bones; then we cut it into strips and made a light solution of salt and water. We left the meat in this solution overnight. This drained out all the blood. We had a small smokehouse, in which we hung the meat. We kept a smoke going under it, made from willows or cottonwood. The air is so dry in the far north in the springtime that it took only a few days to cure it. By slicing some of it thin we could make a full meal on dry meat alone. When one comes into town with dry meat the whole town knows about it in less than an hour. The moose we dried filled two good-sized sacks. We had to hide one sack for ourselves and the other sack we gave away. When the bottom was reached in this sack we told the people it was the end of the dry meat. They believed us and never asked us for more. After that when we ate dry meat we had to lock the door so the neighbors wouldn't see us eating the meat we had held out on them.

In our former trapping years we owned some smart dogs, both Huskies and Malamutes. Their intelligence is about the same. I remember one dog team I owned. They seemed to know trapping better than I did. One January I made a crosscountry trip using only three dogs. I could not drive the dogs ahead of me in going through thick brush—more than three dogs would get tangled up in the brush too easily. I set quite a few marten traps and caught several. The next year I wanted to go back to the same place. I didn't know whether I could make it or not. There was nothing to guide me in the deep snow—no blazes or axe cuts. But I still had my leader, Wolf. Your leader *is* your dog team. If you have a good leader you have a good dog team—if you have a poor leader your dog team is not so hot. I got the dogs on top the riverbank—the same place that we had gone up the former year and hollered "Mush!" to Wolf. He looked like a big grey wolf. He was a full-blooded Husky. I let the dogs take their time. About a half mile back from the river Wolf stopped and looked back at me as if to say, "Here is one of your sets." I walked up to Wolf and dug the snow away using one of my snowshoes as a shovel, and sure enough, I uncovered one of my traps from the year before. I reset the trap and said, "Mush!" again to Wolf. Off he went. The going was slow on account of the deep, loose snow. The dogs had to wallow through it. My place was ahead of the dogs, breaking the trail out with snowshoes. I couldn't do this when I didn't know where to go; everything depended on Wolf to find the trail. Wolf stopped at every trap, whether it was hung up in the willows or had fallen to the ground. The brush and scrubby spruce was so thick the dogs could hardly get through. I had a small toboggan, wedge-shaped, just for getting through brush without having to cut any of it out of the way.

My trap line on the Salmon River, when trapping for mink and lynx, used to take in the sloughs which were frozen up—better than the river itself. One of these sloughs had an old game trail running the full length of it—over a mile long. This trail made a natural place to set lynx snares and traps. Every lynx that hit this trail would follow it. The dog-team trail ran down on the slough ice at the head of the slough. I used to stop the dogs and climb up the twenty-foot slough bank with my gun and snowshoes and follow this slough bank. I used to holler "Mush!" to the dogs to encourage them. It was probably an hour and a half before I reached the mouth of this slough—my dogs would always be waiting for me there. Once it looked as if the swing dogs tried to get the leader to go, because he was lying behind the toboggan facing the wrong way, upstream instead of downstream. The other dogs dragged the leader quite a way, apparently trying to get him to go. When the dogs heard me coming they all started to howl. After I got the leader untangled they went so fast for home that I had a hard time hanging on to the toboggan. We made the three miles home in record time.

One Christmas morning I left Fort Yukon at 3:00 a.m. I intended to make it home for Christmas night. At that time the trail for the upper Porcupine River points followed the river and bars. Nowadays the portage trails are used nearly all the way. The only portage at that time was two miles of a trail north of Fort Yukon leading to the Porcupine River. After reaching the river I stopped the dogs to rearrange my load on my nine-foot toboggan. I spread my lynx bag into the toboggan and crawled into it. Lynx were cheap those days, about four dollars each. I sold them in later years for as high as eighty dollars. A sleeping bag, or robe, made from lynx skins is very warm, which accounts for their being out-of-doors during all temperatures—even down to seventy degrees below zero. Their coat of fur has to be warm to stand it. The rawhide basket on my toboggan made a perfect bed for me to sleep on.

The next portage was twenty miles up the Porcupine River where the trail led up a high bank nearly straight up and down. I felt snug in my toboggan bed, and the temperature was about twenty below—just right for the dogs to travel. I hollered "Mush!" to the leader and away they went; apparently I dozed off to sleep. Four hours later I awoke when the dogs stopped halfway up a steep bank. I unloaded myself from my nice warm bed and gave the dogs a boost up the bank feeling fully refreshed. This portage was about five miles long to the place it hit the river which was called Boxcar because the tiny cabin was shaped like a boxcar. I call that placing great confidence in your dogs. Anything could happen to a person traveling asleep in this manner. The toboggan could tip over and leave you sleeping on the river trail; your dogs could spot game and run away, possibly pulling you into open water or running into a pack of wolves. I never thought of all this at that time. It seems when a person gets older he doesn't take such chances.

I remember how bright the northern lights were that particular night. I fell asleep admiring them. The skies were lighted up like a full moon, and they were all colors of the rainbow. They were making the hissing sound lots of people don't believe the northern lights make. But they definitely do make a hissing sound; I have heard them, not once, but several times, as they shoot across the skies. Being north of the Arctic Circle shouldn't have affected whether they make a noise or not.

At Boxcar I gave the dogs a good feed of king salmon that I had brought along from Fort Yukon, and a few hours rest. There was an old stove in the boxcar cabin. I started a fire in the heater stove and melted some snow for water and coffee. I got the snow from the roof of the cabin where it would be clean enough for making coffee and drinking. I stayed there about four hours. The dogs had rested and I felt fine myself. That sleep I had on the toboggan fully refreshed me. Going up the Salmon River wasn't so easy as traveling up the Porcupine River. I still had forty-five miles to make before reaching home. I left Boxcar at 2:30 P.M. It took me ten hours to make the forty-five miles home. The dogs and I were all tired out. Two-thirds of this distance was made in the dark. There was always the danger of falling into a spring hole

CHAPTER IX — Home at Last

in the night and possibly drowning. The water is always deep where spring holes are.

Once I had a trap line leading from Salmon River to Christian River. This trail crossed a large meadow swamp, more than two miles across. This meadow was covered with short brush grass and "nigger-heads." My trail crossed this swamp and hit the trail where it entered into the brush and timber on the opposite side. I used to have a hard time myself finding where this trail left the swamp. But, as long as Wolf lived he used to hit it on the bull's-eye every time.

When one owns a smart lead dog like Wolf was it is hard to replace him. Money could not have bought Wolf from me. Some dogs are natural-born leaders and easy to train. I used to keep two leaders, exchanging them approximately each week on the trap line. The leader has the hardest time of it; he has to break out the trail for the rear dogs to follow, just as the hardy men broke out the trail for others in the Arctic Circle.

Epilogue to the Second Edition

In 1957, Exposition Press of New York published the journals of James A. Carroll under the title "The First Ten Years in Alaska – Memoirs of a Fort Yukon Trapper 1911 – 1922." These same memoirs were also serialized in 1959 in *Alaska Sportsman* magazine.

The memoirs depicted an adventurous, often humorous, and fascinating period of Carroll's life. His storytelling style was conversational and non-boastful. Carroll was not afraid to tell a story on himself.

In the years following 1922 where the book ends, Carroll continued to trap, hunt, and cook. He also built a Fort Yukon trading post in 1927, which he operated into the 1950s outfitting hundreds of miners, trappers, and local families. He was widely known as a humble and generous man, one who gave cheerfully to those in need.

James A. Carroll wrote on the back of this photo, "My store in the building stage." Carroll opened the store for business in 1927. (Courtesy Carroll family)

James Carroll established his trading post in 1927. Here, Carroll, left, enjoys chatting with a friend. (Courtesy Carroll family)

Fort Yukon circa 1927. See arrow for James A. Carroll's store. Fort Yukon Hotel is in the foreground. (Courtesy Candy Waugaman)

James A. Carroll's son, Tommy, participated in military search and rescue missions via dogsled during WWII. He also won dogsled races while in the service at Gulkana and Livengood, Alaska. (Courtesy Carroll family)

Epilogue

Carroll and his wife, Fannie, raised twelve children, all of whom survived childhood except Mabel who succumbed to Meningitis at the age of twelve. Carroll delighted in sharing stories of his life with his children, telling of the time he walked seventy miles in three days, or the time he endured temperatures of minus seventy-eight degrees.

James A. Carroll's son, Joe, drafted for WWII, is shown here on leave. Joe was stationed at Boeing Field in Seattle. Joe had a reputation for strength from tearing decks of cards and phone books in half. The Carroll family calls it "The Fannie Carroll Grip." (Courtesy Carroll family)

James A. Carroll's son, Richard Carroll Sr., at the Cold Weather Testing Center in 1946 in Fairbanks, Alaska. (Courtesy Carroll family)

James Carroll died at his Fort Yukon home on March 7, 1963 at the age of seventy-two. His beloved wife, Fannie, died of heart failure in the fall of 1980 while residing at the Denali Center in Fairbanks.

At the time of this writing, nearly all the children and grandchildren of James and Fannie Carroll live in northern Alaska, and approximately one-fourth of the population of Fort Yukon is comprised of Carroll family members.

Other books about Fort Yukon

A Schoolteacher in Old Alaska written by Hannah Breece and edited by her great-niece, Jane Jacobs, The memoirs of Hannah Breece.

Bird Girl and the Man Who Followed the Sun by Velma Wallis, Athabaskan Indian legend from the upper Yukon River.

Born on Snowshoes by Evelyn Berglund Shore, Three sisters run a trap line 280 miles beyond Fort Yukon.

Doctor Hap by Clara Heintz Burke, as told to Adele Comandini, The fresh and appealing love story of a dedicated couple in the Alaskan wilderness from 1909 to 1936.

Fort Yukon Trader Three Years in an Alaskan Wilderness by Masten C. Beaver, Northern Commercial Company trader at Fort Yukon 1943 – 1946.

Frank Yasuda and the Chandalar by Ernest N. Wolff, Bits and pieces about Frank and Nevelo Yasuda inspired during the author's trip with Frank in the late 1940s.

Journal Du Yukon 1847 – 48 by Alexander Hunter Murray, Alexander Murray founded Fort Yukon for the Hudson Bay Company in 1847 (French & English).

K'Aiiroondak: Behind the Willows by Richard Martin, Athabascan stories from the Yukon Flats and Porcupine River country (Gwich'in & English).

Living in the Chief's House by Katherine Peter, About the life of the author, the adopted daughter of Indian Chief Esias Loola, and life within in a prosperous family in Fort Yukon, 1926 – 1936.

Neerihiinjik, We Traveled from Place to Place by Johnny Frank and Sarah Haa Googwandak, The Gwich'in stories of Johnny and Sarah Frank (Gwich'in & English).

Raising Ourselves by Velma Wallis, A Gwich'in coming-of-age story from the Yukon River.

The Arctic Forests by Michael Henry Mason, The author visits Fort Yukon in the early 1920s.

Two Old Women An Alaska Legend of Betrayal Courage and Survival by Velma Wallis, Athabaskan Indian legend from the upper Yukon River.

Voyages on the Yukon and its Tributaries by Hudson Stuck, A narrative of summer travel in the Interior of Alaska.

Wolf Smeller by Clara Childs MacKenzie, A biography of John Fredson.

These fine books, and
thousands more, available at:
Alaskana Books
564 S. Denali Street
Palmer, Alaska 99645
E-mail: alaskanabooks@alaska.com